THE ADVANCED COURSE
IN
PERSONAL MAGNETISM

THE SECRETS OF
MENTAL FASCINATION

BY

THERON Q. DUMONT

Instructor in the Art and Science of Personal
Magnetism, Paris, France

Develops Your Personality and the Personal
Power that Sways and Compels and
Gives You a Powerful Influence
Over the Minds of Others

DOMINO PUBLISHING COMPANY
PHILADELPHIA, PA.

ENGLISH REPRESENTATIVES:

L. N. FOWLER & CO.

7. IMPERIAL ARCADE, LUDGATE CIRCUS,
LONDON, ENGLAND.

COPYRIGHT 1917
CHICAGO, ILL.

CONTENTS

Chapter I. Advanced Course in Personal Magnetism..11
Everyone possesses a tremendous power. The charming personality. The law that evens up. Magnetic power enables the business man to secure credit. The qualities of strong personality. The doors are open to those with the pleasing manner. The Power of Magnetism in business. The hidden beauties of our nature. Personal Magnetism is not a subtle force possessed by but few, but is latent power that can be developed. Money spent in making yourself magnetic is well spent. A magnetic woman can change a man wonderfully. Why men and women become tiresome to each other after marriage.

Chapter II. The Secret of Being Naturally Magnetic..20
A startling chapter. Man's most valuable asset. The blessing of friendship. You can become magnetic and attract friends and success. The Magi, the wise men of the East could influence people. The priests of ancient Egypt were adepts in the art and practice of magnetism. How to gain control of others. The dual consciousness of subjective self. The only difference between the manual laborer and the mental giant. The inner mind, the real secret of success. The deeper and greater personality. Why we think and act as we do. The correct solution of your problems. What a strong and determined will, will do. How to become more magnetic. How to increase and improve your power.

Chapter III. The Development of Your Magnetic Power ..29
How to attract magnetism. Rules for developing a magnetic personality. The unfolding of your latent powers. How to make yourself more proficient. The only things worth while. Why selfishness is unmagnetic. The art of being agreeable. The mistake of many wives. How to become well liked. Acquiring attractive manners. The man that is well fitted to control others.

Chapter IV. Some Pointed Helps and Golden Laws....41
How to develop a magnetic voice. Power of Conversation. Traits the majority of mankind are heir to. Your undesirable qualities that add nothing to your comfort or welfare. The desirable qualities that are a great help to you.

Chapter V. How to Develop Physical Power..........59
What makes a winning man. Qualities of a magnetic will. The unseen currents around you. How to impress others. How wives have made great men out of their husbands. What you must do to become more magnetic. The law of demand. The universal supply. Suggestions in character building. Your disposition is written in your countenance. How to make yourself good-looking. Helps in controlling others. Develop physical magnetic power.

Chapter VI. How to Develop Magnetism by Self-Suggestion ..63
You can accomplish what you believe. What the men that achieve great things possess that the ordinary man does not. The man that

succeeds. How to double your power and multiply your ability to do anything you have the longing to do. How we can attract the things we want. A method for building a timid character into a strong bold one. The power of personal suggestion. You have a place in the world, fill it like a man.

Chapter VII. How to Use Your Personality to Win the Affection of the Opposite Sex 76

The development of your fascinating qualities. Traits that men admire in each other. The capabilities of a real lover. What men and women are fond of. The wife that holds the husband's affections. The wonderful magic in love. The desirable men and women. The reading of a person's character and purpose at a glance.

Chapter VIII. The Important Part of Magnetism in Love, Courtship and Marriage 83

Love, the regenerating process of nature. The cause of more unhappy marriages than happy ones. Advice to those about to marry. Nature's plan. The magnetic person has plenty of chances to marry. The control of one mind over another. Some secrets of influencing others. A person may be under the complete control of another's thought without them knowing it. How our magnetic force is greatly diminished. Why people get old before they should. Jesus replenished his supply of magnetic force. How you can accomplish a great deal more work. The exchange of magnetic force is mutually helpful. Control the elements of success.

Chapter IX. What Constitutes a Pleasing Personality .. 92

What constitutes a pleasing personality. The important part personal appearance plays in your future. Personality a big factor in business. How to improve your personality. The value of first impression. One of the drawbacks of men seeking advancement in the business world. The value of politeness. One of the biggest stumbling blocks to the development of personal magnetism. Qualities of a gentleman. What a different world we could make of this. The art of pleasing is the art of rising in the world. The voice and the eye cannot lie. How to avoid shyness which repels friends instead of attracting them. A simple method how the author transformed an awkward young woman into a graceful, attractive well poised girl.

Chapter X. The Wonderful Power Within Us 118

Your latent powers are capable of wonderful things. The control of your mental forces. Weak people are controlled by forces. Strong people control the forces. The adept character reader. What you must do to be popular and magnetic.

Chapter XI. Vital Magnetism 126

Why you must be capable of having magnetic power before nature endows you with it. Methods of cultivating magnetism. Making your hands and feet magnetic.

Chapter XII. The Law of Magnetic Thought Attraction 133

The kind of people that we like. How to be happy all the time. The value of cheerfulness. You can make a wonderful difference in yourself. Make yourself positive, vital, magnetic. Realize your highest dream. You have no weakness unless you think you have. Concentration can bring you what you wish. Make yourself a magnet and draw what you wish.

Chapter XIII. Why We Are Judged by Our Thoughts. 144

How we form our opinions of others. What we think about we become like. Your secrets can be read. Be careful of your thoughts. You have no limitations. What your success in life depends on. Making yourself clever and capable.

Contents

Chapter XIV. Magnetic Character Building.........151
What our great philosophers tell us. Your weakness and why they show more prominently than your strong qualities. How to cultivate magnetic beauty. Control the gate to your brain. Putting forth your best effort.

Chapter XV. The Secret of Personal Power..........157
Those that find life interesting. How to prevent anger. The law of just compensation. The opinions of others. Great people are totally unconscious of others' opinions. Those that stand like a stone wall and are not affected by the storm.

Chapter XVI. How to Cultivate Success.............163
Your success depends on your poised power. Mental independence brings personal power. Never cramp your intellect. The reward that comes with power. Discover yourself and you will be ashamed of your doubts. The lower down a man is the more likely he is to rise. The force that drives men to reach any goal. Why he has dominion over earth. You have enough power within you to overcome any outside power.

Chapter XVII. Qualities That Will Make You Successful 171
Everyone needs a stimulant at times to do their best work. The plan of a big advertising firm. The important part your associates make in your life. Don't dull your intellect and powers. The men that you are the better for knowing. Watch your actions and keep them up to the standard. The chances for the young man that applies himself are greater than ever. Experience a dear school but the only one fools will attend. The secret of a man that raised himself from $700.00 per year to $50,000 a year.

Chapter XVIII. How to Protect Yourself Against Injurious Thought Attraction 178
Expectancy is a powerful magnet. The power of sympathy. How it is possible to know what another is doing though a long way apart. You have the power to transmit thoughts. The key that unlocks the door of every heart. How to broaden your mind and understand all kinds of people. The motto of the happiest man in the world. An occult law. How to become wise.

Chapter XIX. How to Make Yourself a Great Power in the World 186
Your divine creative spirit. How Napoleon and Cromwell turned failure into success by the force of their mental powers. The books that inspire noble thoughts and lofty ideas. Don't be a doubter. There is always a way out of difficulties. Knowledge gives you power. Your most powerful weapons in dealing with others. You have two pair of eyes. Your eyes reveal truth or falsehood.

Chapter XX. Why Personal Magnetism Prevents Disease 197
Magnetism promotes health, ambition and cheerfulness. Recharging yourself with vital force. The two qualities essential in learning the power of influence. Special exercises that will help you secure love, and retain it. Making the whole body magnetic. If you follow the instructions of this book for six months you will hardly recognize yourself. How to develop a compelling magnetic gaze. Attractive powers are not inherited. The object of the author's courses are to make you a positive, conscious and intelligent centre.

Chapter XXI. A Formula for Creating Happiness.... 122

    The secret of happiness and an antidote for worry poison. It will cure you of most of your troubles. How you destroy your poise. A tonic that for many ailments is better than a doctor. An illustration that will show you how valuable it is. How to fortify yourself for a rainy day. How you weaken your powers and make yourself less certain.

Chapter XXII. The Man and Woman Thou Were Intended to Be 217

    The blessed omniscience state. You are not the product of chance. The coming man of the higher civilization. The magnet that will increase your prosperity and happiness. The new man will be healthy in soul, mind and body. You are one of the tiny spokes of a huge wheel that is gradually turning to perfection. How to make yourself happier and feel far better. The author's final word. **Special instruction and advice.**

FOREWORD.

In my previous work, "The Art and Science of Personal Magnetism," I gave the principles and elementary rules for developing personal power and influence.

In this work the subject will be continued, although it will be a complete work in itself, as no doubt many will secure this volume before they know there is a previous work. Those who have read "The Art and Science of Personal Magnetism" will better understand these advanced lessons, because we have repeated very little of the matter contained in the former work.

I will not go over old beaten paths, or indulge in abstruse terms, but will give in a condensed non-mystical style all I have been able to learn of this wonderful power, from the latest and ablest teachings on the subject.

As in my previous work, all technical terms have been avoided, so that even the layman may readily understand. Some time ago one of my students asked me why I had met with such remarkable success in my teachings, and my reply was that, after many years of study, I have succeeded in working out a united system along purely scientific lines, which is practical and appeals to reason. In a short time the faculty of intuition is developed, and as one secret is unfolded the key to another is placed in the hands of the student.

For years, at a great expense, I bought course after course of instruction. I studied everything I could get hold of relating to the secrets of personal magnetism, and success. Many of these instructions were guaranteed to accomplish wonders, and I earnestly waded through their abstruse terminology, hoping to find the key.

I found that many others had pursued the same course that I had, and that the results were unsatisfactory. Then I concluded that something must be wrong.

In time I realized that the private lessons, which were expensive, were got out as a money-making proposition. They were not concerned with the effort to give students value received; to make the work concise and readily understood and applied, but merely to confuse.

Naturally this method of procedure could not attract success. The foundation upon which they built was wrong.

In my search I found some grand truths. I gained a little from each one, and thus I have been able to put into a condensed and perfectly simple form all the secrets which I have discovered and proved, so that I am confident that the student who earnestly learns and applies these straightforward and practical methods will be able to win success. They have been worth thousands of dollars to others, and will be, I know, to you.

What I am giving you in this volume has been personally tried and proved. From my long experience I am certain that if you study the lessons and practice them, you can become magnetic and make good. If you have not met with the success you desire; if you have not been able to carry out your plans, just get busy and read my lessons carefully. Through your higher self you can come in contact with tremendous forces and you can master the secret of the Omnipotence of Natural Law.

One thing I must impress upon you, and that is that you must always use your forces wisely. You must never attempt to use your power to further anything that is dishonest.

If you should thus attempt to misuse the power you acquire, I can assure you with absolute certainty that you will draw unto yourself penalties in exact proportion to the wrongs you have done.

CHAPTER I.

THE ADVANCED COURSE IN PERSONAL MAGNETISM.

Everyone possesses a certain magnetism which eludes the photographer; which the painter cannot reproduce; which cannot be chiseled by the sculptor.

Although it cannot be seen, nor really described, personal magnetism is a tremendous power. It is a very important force in life, and those who possess it will meet with the most happiness and attain the greatest success in life.

Many poor boys and girls have envied the rich youth who does not have to go out in the world and earn a living, when all the time they may have in their possession a magic power, which, if properly developed, will bring them greater honor and success than that of the rich man's money.

The man or woman with a charming personality is wanted everywhere, while the millionaire cad will not be tolerated, except to be made a joke of. The man with a magnetic personality, with a very limited capital, is far more likely to succeed in business than is the rich man who has an unprepossessing personality.

There is a law back of all creation which evens up accounts. No one is given all the advantages.

There are many young men who have started in business with almost nothing in a financial way, but who possessed a strong, clear magnetic power which enabled them to secure credit. Jobbing houses, wholesale dealers and business enterprises of all kinds are willing to take a chance with a pleasing young man of promise.

We meet persons who are well-groomed, persons who have agreeable manners and are attractive, persons who possess a great deal of magnetism, but we seldom meet one who possesses all these.

In this work I shall show you how to cultivate all of them. When you have acquired all these qualities, you will possess a talisman which will open all doors to you.

The person with pleasing manners is seldom refused any request. They have such an attractive way with them that you feel it is a pleasure to do something for them.

You have all met such persons. I know a salesman who can sell almost anything. Here is his secret: He interests himself in you at once, and makes you feel that he is doing you an especial favor by letting you buy his goods. He is so attractive and his manners are so ingratiating that you admire the man. He brushes aside all formality and in a short time after you meet him you exchange confidences as though you had been life-long friends. He is so different from other salesmen that you feel it would be a pleasure to do business with him. The chances are that he may sell you something that you do not want, but at the same time he will give you the impression that he would rather lose his right arm than to permit you to take anything you do not need.

THE POWER OF MAGNETISM IN BUSINESS.

The promoter knows the value of personal magnetism and does everything in his power to develop that latent force in himself. The business man is now beginning to realize its power and is looking for employees who **have** a winning personality, those who may become magnetic by a little training.

Few persons are proof against a persuasive influence.

In every walk of life we are constantly swayed by the charm and the magnetic personality of some whom we meet. Those who are the most fascinating achieve the greatest successes.

We are helped and strengthened by mere association with magnetic persons, those whose character is charming and attractive.

They seem to have the power to lift you upward. You feel broader and bigger in every direction. They make you realize possibilities you never dreamed of. They awaken in you a new power; you are inspired with a new ambition. They make you want to do things. The best that is in your nature is drawn out; you have new impulses and aspirations thrilling you.

It has always been a moot question as to just what is the greatest blessing in life. We all have different ideas. Most persons think that the acquisition of money is all-important. But it is my conviction that the **greatest** achievement we need hope for is to raise ourselves to our highest possibilities; to call out the hidden beauties of our nature, and to become attractive and helpful to humanity instead of repellent and unsympathetic.

For the benefit of those who have not read my previous work, "Personal Magnetism," I want to state that I have taken students who appeared to be totally lacking in magnetism, and in a short time have developed them into the most charming men and women.

Personal magnetism is not a subtle force possessed by only a few, but it is a latent power which may be developed by every one of fair intelligence. By persistent cultivation any one may possess attractive qualities.

All that I received in the many courses of instruction which I studied, and much more, will be contained in my books, set forth in the simplest and most lucid language, so that there is no reason why you should not become agreeable and pleasing to every one with whom you come in contact.

Your love nature will be developed and your sympathy will be broader; you will become kinder, gentler and more interesting to your fellow-beings. Any amount of time and money expended in the work of making yourself magnetic will be well spent and will pay you well.

It would be difficult to estimate the number of persons who have voiced the following wish: "If I could only attract friends; how I wish I could become popular and magnetic."

Those lonely, discouraged souls who, through lack of magnetism, are left out in the cold world, could readily acquire this priceless possession if they would but resolutely apply themselves.

It would be well for young persons about to enter mercantile life, and also for others who have thus far failed of the success they had hoped for, to consider this opinion expressed by experienced and highly successful men: "A pleasing personality is not only a valuable asset, but is almost indispensable to the unknown aspirant for favor. There are doubtless thousands of men and women in this and other cities who have turned away disappointed, after making application for a position as salesman, without having the slightest idea why they failed to get a chance to work. They have been competent to do the work they applied for, yet, because of some carelessness in dress or sloppiness of manner, they made a poor impression on the man who appraised them, and thus they lost the opportunity to show what they could do. If they had considered beforehand that strangers are bound to judge their capabilities by their appearance, they might have improved that appearance enough to have turned the scales in their favor.''

There are a few persons who are born magnetic, but the power is usually acquired by cultivation. Those who do have it are students of human nature. Magnetism is acquired by studying magnetic people and imitating their ways. Magnetic people impart some of their power to those who associate with them. It is a law of nature that whatsoever you give out will return to you with interest. The magnetism you transfer to others will return to you with added power.

You will find the man that does not associate with women has no magnetism. Women usually have more magnetism than men. Man draws much of his magnetism from woman. It is through her influence that he becomes refined and interesting. Many coarse, stupid and uninteresting men have changed themselves in a short time after having become interested in some woman.

A woman of strong magnetism can accomplish wonders with a man in a short time. By a strange "irony of fate," man, after being taught by woman, often turns and uses the power he has thus acquired to control her and compel her to do his bidding.

He is able to do this because his training generally develops a stronger will. Every woman has her master. She is a willing captive to the right man, and when she is thus captivated she is a wonderfully changed person. She loses much of her individuality and what is her husband's will and desire becomes hers. This is all wrong.

You should never subject your own individuality to another's. You should seek to preserve your individuality and to develop and improve yourself. If you admire some one for their attractive qualities, add those same qualities to your own, but retain your individuality.

Many women after marriage so change themselves that they become a mere tool of their husband's will. Thus they become monotonous and tiresome to the husband whom they seek to please.

Nor is this state of affairs entirely one-sided. Many a man loses his finer qualities when he falls under the domination of an inferior woman, and instead of developing along the lines of his higher nature, he permits himself to be influenced along the wrong lines. It is important that we each and all preserve our own higher instincts and develop strength and magnetism without seeking to subject another to our will.

CHAPTER II.

THE SECRET OF BEING NATURALLY MAGNETIC.

What I am going to state in this chapter is startling, but after a careful study you will find that the laws governing phenomena are natural laws.

If I were asked what I consider man's most valuable asset, I would unhesitatingly answer "Personal Magnetism."

Personal Magnetism will bring happiness to the person possessing it. A man may sometimes succeed in acquiring worldly goods without the aid of this wonderful force, but he will miss the greatest happiness in life, namely, the love of family and friends. Work will no doubt have given him a certain satisfaction, but when the twilight of old age has been reached dollars are a cold substitute for love and friendship.

This work will, I trust, be read by a great many young men and women who are just starting in life, and perhaps by many others whose life has thus far been a failure.

To these I want to say that you should read this work carefully and then put into practice the teachings I give you. You can, if you will, become magnetic and attract to yourself friends and success.

Of all the millions of people there are not two who have the same kind of magnetism; each person is distinctive; his magnetism is entirely his own. We know this from the fact that a bloodhound can follow a person through a crowded street guided by the trail of the person's magnetism.

The secret of being naturally magnetic is to develop your love nature so that you think love of all the world; you must possess self-control; you must at all times be above petty meanness; above irritability; above resentment and malice and gossip, and all the weaknesses which so many permit to interfere with their highest and best development.

You must develop your generosity and your sympathies, and be ever ready to give a helping hand to the needy.

Knowledge is self-conscious power, and by thinking how you can add to your personal influence you can do so.

Many volumes could be written from ancient records demonstrating how The Magi, the wise men of the East, could influence the people. History proves the fact of the possession by the few of the secret of this power, and as far back as we have any records we find evidences of the secret of fascination and its marvelous effect upon the world.

In fact, it would seem that in the earlier stages of our evolution fascination and the laws governing the use and the development of personal magnetism were known and practiced more than now.

Man has no limitations other than those which are self-imposed. All things come to him who strives with an earnest, unfaltering will.

No one can study the records of ancient Egypt and the wonders which their priests performed without realizing that they were adepts in the art and practices of magnetism.

HOW TO GAIN CONTROL OF OTHERS.

You can never control another unless he or she is negative to you. If you have studied my previous work, you should be able to make yourself positive to others, at will, but as you may meet some persons who are extremely positive, I will first show you how to make them negative.

Scientists are giving more attention these days to the subject of "dual consciousness" than ever before in our modern research.

Nevertheless, according to the ancient occult teachings they are still only at the outer fringe as it were of that vast field of human activity.

A more careful and extended study of the areas of sub-conscious activities is bound to reveal some wonderful secrets of our inner selves. Every one has a deeper and a greater personality than that which is evident to the casual observer.

The only difference between the manual laborer and the mental giant is a matter of brain development. In the man who works only with his hands the latent powers of the subjective self are still dormant. In the brain worker these faculties have been aroused.

We speak of "doing things," but before we commence to do, our inner self must marshall the facts and co-ordinate them for our hands to work out. As yet comparatively little is known of this inner mind. but we are commencing to realize what wonders it can do for us. The inner mind is what makes a man. It is the real secret of success. It is very important, therefore, that you come into close relationship with this inner mind, both in yourself and in others. You can trust your inner mind *to solve* the most difficult problems and guard you against defects.

Experience is a great teacher and your experiences are all stored away in your subconscious mind, to be called upon at your will.

We think and act as we do because of past experiences. Of course we do not remember all of the things that have happened, but they have all left an impression upon the inner self.

Ask a successful business man why it is that his plans turn out right and he will say, "I just used common sense." J. D. Morrell says: "Common sense is nothing but a substratum of experiences out of which our judgments flow while the experiences themselves are hidden away in the unconscious depths of our intellectual nature."

In "The Diseases of Memory," by Ribot, a curious case is reported. A business man in Boston, Mass., having an important question under consideration, had for the time given up the solution of it as too much for him. But he was conscious of an action going on in his brain which was so unusual and painful as to excite his apprehension that he might be threatened with palsy or something of that sort.

After some hours of this uneasiness, his perplexity was all at once cleared up by the natural solution of his doubts regarding the problem which he had given up—worked out, as he believed, in that obscure and troubled interval.

In the instance above quoted the man was vaguely cognizant of the work going on in his brain without having distinct consciousness of it.

"You have but to look to your deeper self for your best problems and your best solutions, some of which come because you seek them, but many of which come when you are not seeking them."

You can educate the inner mind to such an extent that you can depend upon it for help, and it will do as you suggest.

To a person who has never read along this line of thought some of these teachings may sound mysterious, but I can assure you that what I am telling you is actually going on all the time, only your outer, or objective, mind is unconscious of it.

Your deeper self always decides your most difficult problems. Sometimes you are conscious of thinking deeply, but more often this inner self sees the matter under consideration without the fact ever rising to your outer consciousness.

By following my instructions you will be able to educate this inner mind and to consciously depend upon it for help.

We know that if a man starts out with the conviction that he can succeed, he will be much more likely to attain success than will the one who doubts his ability; so it is with this inner mind. Know that you can succeed in educating this mind, and you will soon discover that your commands will be obeyed.

Instead of meekly asking for the solution of a problem, you should demand it, and you will get it. To a certain extent you have always done this, but the point is to do it consciously, with power and with confidence. Thus your power will increase.

Man has hitherto placed limitations to what he might do.

Gradually we are lifting this barrier to accomplishment, because we have begun to realize that everything gives way before the force of a strong and determined will.

In our schools the mind is trained, but not the will. The will is higher than the mind and should control the mind just as the mind controls the body.

HOW TO BECOME MORE MAGNETIC.

The first thing I want to impress upon you is the importance of faith and courage. According to your faith and your courage to persevere will be your increase in power.

Impress this fact firmly upon your mind: If I care enough for results I will attain them.

If it is really your desire to be magnetic you can be; if you want to be rich you can be; but, before you will be, you must decide just what you want to be and concentrate upon it. The trouble with the majority of persons is that they try to do too many things, and do not concentrate enough. If you want anything and want it above all other things, you will get it.

Cultivate assurance and divorce yourself from doubt. Whatever you undertake, make up your mind to be successful. Every person has some desirable qualities, and also some undesirable ones. Eliminate the undesirable ones and build up the desirable. By careful study of yourself you will be astonished to find how much you can improve your power.

CHAPTER III.

THE DEVELOPMENT OF YOUR MAGNETIC POWER.

Every person has already all the magnetic power that he or she will ever have any use for. What you need to know is that you have it, and to be able to draw upon your storage-keeper—the inner self.

You cannot make yourself magnetic by merely reading all the books in the world on the subject. No one can impart power to you, because no one can create power for you. All power comes from within. It does not take any more effort to be magnetic than not to be. To unfold latent power is very simple. When once you realize that you have power, you have confidence in your efforts, and this power irradiates your whole being. If the irradiation is weak, your power will be weak, but if the irradiation is strong, your power will be proportionately strong, and you may become invincible. We never know what we can do until we have something to spur us on to achievement.

You can develop magnetic power by thinking and acting and "imagining" that you are more magnetic. Thus you become a human dynamo and you attract magnetism. Once you start the subconscious self thinking that you are magnetic, your whole personality will respond and you will develop unlimited power.

An Oriental saying is: "There is no limit to the knowing of the self that knows." To develop yourself so that you will come in close contact with that self that knows is a coming science. We will not take up this subject in this work, but will in a later work, which will be called "Seership: The Science of Knowing the Future."

You will find the person with a winning personality has something behind him. A weak person is never magnetic. I do not mean physical strength. Physical power is not attractive power. But physical power may be transmuted into genuine magnetic ability. Physical power is merely the ammunition which may be used for other ends.

RULES FOR DEVELOPING A MAGNETIC PERSONALITY.

The rules for developing a powerful personality are many, but after a long study I have gradually weeded out the least important ones and, whereas some of the rules were in the beginning rather difficult to learn, I have them now worked out into a simple and concise form. If you will apply the following instructions you will soon commence to unfold your latent powers:

RULE I.

Believe in yourself; that you have all the power you wish.

You can readily see the object of this. If you think you have no power and that you are weak, this thought will not develop any power, but, on the other hand, if you think you are powerful and start the subconscious self to work, you are going to develop power. Your negative state is your weakness and your positive state is your strength. Thus you now realize the advantage of keeping the idea before you that you are developing magnetic power. Commit this to memory:

"I am power. I am equal to anything that

may come up in my life. I will unfold personal power. I am creating within real dynamic personal power."

The more power you think you possess the more actual power you will unfold and it will be the kind that will bring you real dividends.

RULE II.

Make up your mind to believe in yourself. Let your whole manner denote action; confidence, courage and the ability to do things.

You hear men say, "He will make good." What is there about the man in question that gives them this opinion of him? It is his manner. He believes in himself.

Suggest to yourself: "I am power. My manner and my personal atmosphere (a sphere of space around me) vigorously expresses activity and courage and confident personal power. It is my way. It is my attitude. I have enough to win confidence and respect."

RULE III.

You can never hope to increase your personal magnetism until you first familiarize yourself with the tools pertaining to the **power**. The tools are those of your own organism. You will find that at times you are very unmanageable. The mechanic becomes trained in the use of the tools of his trade, but few persons ever become skilled in handling the tools which nature has given them within their own mind.

The physical body represents the storage-house. Your success in life depends upon your ability to handle this power. Watching others is of immense benefit. If they display qualities that would be of value to you, make a mental note of them. Many persons do things quite unconsciously. You may do the very things which when seen in others horrifies you. What we do ourselves never looks quite as bad to us. The only way you can improve yourself is to watch yourself closely each day. Be determined that you are going to correct some fault, or do something better than you have ever done it before. You will see that this is very easy of accomplishment. In a year's time you will have made three hundred and sixty-five improvements. At this rate you would assuredly have become proficient. It is a good idea to have a little notebook. Every time you do something which you know you should not have done, make a note of it. The next time you do it, the thought will immediately suggest itself, and the chances are you will not repeat the undesirable act.

You will find it a great help to write down each night before going to bed, the improvement you have made. This will impress it upon the subconscious mind and it will become a part of you. It will be just as easy to do this as anything else. The suggestion for the rule: Be ever ready to change for the better. Keep your mind receptive to the good and close it to the bad. The law of good is wonderfully magnetic, and it is deplorable that there are so few who realize: "The right thing and the true thing are the only things worth while."

RULE IV.

Be ever careful what you say. Never take for granted that it is the truth because some one told you so. They may have added a little to the original story; most persons do; or they may have been misinformed. If the latter is the case, you may be placed in a very embarrassing position and are liable to have a painful or unpleasant experience.

Unless you are convinced that something which has been told you is really the truth, do not think of repeating it as the truth. If you do repeat it, give the authority it came from. But, even though it be the truth, unless you are ready to say it to the person's face, do not say it at all. The square man will never state facts other than as they are.

RULE V.

You will never find a selfish man with a winning personality. Self-interest is very necessary, but deliberate selfishness is very unmagnetic. Sometimes we meet persons whom we think we would like very much, but when acquaintance develops the fact that they are selfish, we lose our respect for them and we turn from them in dislike.

One of the greatest assets in life is friendship. A man may have all the money he wants, but unless he has real friends he will not be happy.

RULE VI.

Tact is a simple little word, but it has a wonderful meaning. The dictionary defines it "ready power of appreciation and doing what is required by circumstances." If husbands and wives would exercise the same tact after marriage as they did before, there would be fewer unhappy marriages.

THE ART OF BEING AGREEABLE.

You can make yourself so agreeable that no one can be disagreeable to you. "When I was first married I came home several times and found my wife all out of sorts. The first time it was because the maid had let the roast burn, but instead of getting angry as she was, I laughed and joked about the 'awful calamity.' I inquired if there wasn't something else that we could eat. We soon discovered that we could get along very nicely without the roast. The next time I found my wife out of sorts the maid had left without so much as saying she was going. When my wife went out to see that everything was in readiness for six o'clock dinner, there was neither dinner nor maid. She immediately worked herself into a bad temper. Again I gave her the laugh, and in less than a half hour we were enjoying a good meal. After these two experiences my wife realized how useless it was to lose her temper and permit herself to be upset over trifling things, or over big ones. It does no good and does much harm. The weak person loses his temper easily. The strong one controls his. Don't show your weakness by constantly losing your temper.

Many married people would live happy lives if they would learn to control their temper. And it is generally the trifling things that cause trouble. Avoid dwelling upon trifles and you will do away with most of your troubles.

RULE VII.

If you want to be well liked it is necessary to become what is called "a good mixer." This comes naturally to some, but with the majority it must be cultivated. As there are all kinds and classes of people, it is necessary for you to be able to adapt yourself to all sorts and conditions of persons. A good mixer has to be willing to meet the other fellow's tastes and desires, not as a condescension, but as a great pleasure. If he meets some one with a "hobby," he interests himself for the time being at least in the same thing. A great deal of diplomacy must be exercised by the one who is ambitious to develop a winning personality.

RULE VIII.

Have ever an open mind. Do not try to convert others to your way of thinking or acting. Avoid all interference with another's tastes, beliefs, political affiliations, recreations, business affairs, etc. I do not mean that you must go out of your way to do things which are objectionable to you when you do not have to, but it is well to "do as the Romans do when you are in Rome."

It is just as easy to refrain from expressing your dislikes as it is to express them, and you should always remember the importance of tact.

RULE IX.

An attractive manner is assured if we will avoid the following: Sarcasm, impertinence, ridicule, hot temper, profanity, roughness, brutality, vulgarity, a loud voice, and grouchiness. Almost every one has some of these qualities. How many have you? You may have more than you think. It is possible to **exchange** every one of them for their opposite. If you are talking to some one who gets angry, if you will but control your temper his anger will cease. Make a point of cultivating the exact opposites of these traits and it will help you wonderfully to develop an attractive manner. Sometimes, though not often, we hear said of a person, "He is genuine." Men and women of "genuine" qualities are everywhere in demand, and will be more and more so in the business world. Genuineness develops magnetic qualities of the very highest kind. When you meet a person of this genuine stamp, you know at once that you may trust him or her.

If you will observe carefully, you will note the fact that the genuine person does not turn his eyes away when conversing with you. His manner is free and natural, simple and unaffected.

Take this suggestion into your mind and act upon it: "I am just what I appear to be, and no man can say I am not."

RULE X.

There is nothing so detrimental to the development of a magnetic personality as lack of self-control. It is not so much what you say as how you say it that counts. A man who loses his temper loses his balance and does not weigh his words. The result is that he arouses anger and resentment in others as well as himself and defeats his own efforts.

The man who keeps himself under control commands our respect and confidence. Such a man is well fitted to control others. In the morning before you start out to your daily work or play, as the case may be, say to yourself, "No matter what happens today I am going to remain calm and not lose my temper." Before going to bed at night think over your day's actions and see if you have kept your word. If you have, make a firm resolution to continue to do so tomorrow and the next day and all the days to come. You can do it, and you will find that it will pay you well besides helping you to develop magnetism.

RULE XI.

You cannot expect to have a winning personality unless you have an eye that is not afraid to look straight at a person. Look into the mirror and think of something pleasant. Note the effect in your eyes. Practice in this way will give you a magnetic eye.

CHAPTER IV.

SOME POINTED HELPS AND GOLDEN LAWS.

HOW TO DEVELOP A MAGNETIC VOICE.

Many times we see persons who we think will be fascinating, but as soon as they speak we change our opinion. A well-trained voice is a big asset and one that is well worth cultivating. Few persons are naturally gifted with a good voice. All need more or less training. The more variation you can put into your speech the better. A man with a well-trained voice can use an infinite variety of ways in which to express his thoughts.

The first thing you should aim to do is to speak with a pure tone. You should put every bit of your breath into your voice. A voice thus controlled and managed is more agreeable to listen to and requires less effort. If your thought is pure, there is no reason why your tone should not also be pure.

POWER OF CONVERSATION.

You cannot be too particular how you express yourself. Nothing betrays the character of a man so quickly as his conversation.

"The quality of his voice, his use of words, his ability to put his ideas into forceful and effective speech all disclose his breeding and education."

The ability to be able to speak interestingly is very necessary if you want to be entertaining. You must be able to describe vividly an interesting experience or tell a story graphically.

The person who wants to become magnetic must be careful not to use loose or incorrect language. Be ever on your guard. Use the utmost care at all times in your daily conversation. Speak every word clearly and correctly. Use a low tone, as low tones are purer. Never use slipshod language or a light pitch tone. Don't try to speak too fast, and open your mouth when you speak.

When you shake hands put some feeling into it. Your grasp should be firm, but never use a crushing grip. Put your whole hand squarely into that of the one you are greeting.

Be careful of your carriage, and stand and sit correctly. Stand squarely on your feet. When you sit down see that your feet are on the floor and not twisted around your ankle, or around the leg of a chair.

The next time you are in a car or a restaurant, notice how the majority of persons have their feet twisted. See that you do not do this. It looks very bad.

Don't try to talk too much. Be a good listener. If you want to make a good impression upon another, get him to lead the conversation, and have him tell you all about himself. Most persons like to talk of themselves. This is a bad thing to do, and don't you be guilty of it.

All of the above traits are the inheritance of the majority of mankind, in a greater or lesser degree, as the case may be. But you can improve yourself to an unlimited extent if you try, and this I know you wish to do or you would not be reading this book.

Remember that the magnetic person is the well rounded individual. All of your good points serve to make you magnetic. All your faults combine to make you unmagnetic. You may wear a mask for a time and not be detected, but not forever. Your true nature will reveal itself.

What you want to do is to change those qualities hidden by the mask into desirable ones and then you will have nothing to conceal. You will have nothing to hide and you will act freely and naturally.

Undesirable qualities usually hidden by a mask.	Desirable qualities that should be cultivated.
Indifference	Friendliness.
Don't be weak	but vigorous
Don't be nervous	but calm
Don't be fickle	but even
Don't be slow	but quick
Don't be uncertain	but decisive
Don't be repelling	but agreeable
Don't be doubtful	but confident
Don't be fearful	but courageous
Don't be unreliable	but faithful
Don't be tricky	but honorable
These will do you no good, and add nothing to your comfort or welfare.	These will help you in a great many ways.

Take an inventory of yourself. How many of the characteristics tabulated above on the left hand side, have you? Have these ever helped you any? Haven't they caused you a lot of trouble and harm? Wouldn't you be better off with those on the right hand side? Certainly you would. Then why not cultivate them? They will pay you well. It is qualities such as these that will make you magnetic.

The world needs noble characters. Men and women who will stand for what is right, first, last and all the time. Men and women whom money can not buy. There is a big opportunity awaiting those who can qualify.

Make it a point to be agreeable to all with whom you come in contact, and not just to a few persons upon whom you desire to make a favorable impression. Try to win the respect, friendship, and confidence of every person with whom you come in contact. And to do this you must deserve it. Don't be agreeable merely because you want to impress a person, but because you feel kindly toward all.

Do not use slang or profanity. Be careful to make your first impression an agreeable one. Do not appear embarrassed. Don't permit yourself to be annoyed by small things. No matter what your occupation, feel that you are just as good as any one. No work can degrade a man but a man can degrade a position.

If you feel that you are superior to some one else, never show it.

If you will keep cool you will be the master of the situation no matter what may come up.

Be simple and avoid all false notions.

A good reputation is of priceless value. Do nothing to mar it.

We should not judge a man by his clothes, but we do nevertheless. You must not only be all right but you must look all right.

No man or woman wants to meet those for whom they especially care, when they are looking badly. This is inborn.

Take note of some person you know, the next time you see him particularly well groomed. His manner will be quite different from the one he assumes when he has on old clothes.

It is not always the person who expends the most money, who is the best dressed. It requires tact in selecting clothes and in taking care of them.

The following points should be taken into consideration:

Don't try to wear something which does not look well on you, just because it happens to be the latest style.

Study what personally suits you.

Don't dress so as to attract comment, and criticism.

Dress in harmony with your surroundings.

THE WINNING MAN.

is always master of himself and when you are master of yourself you are also master of others, because you are a man of power.

The real instruments of magnetism are:

Making the best of yourself.

Strong resolution.

Decision.

Preserving your energy.

Unwavering persistence.

In studying these lessons keep this ever before you:

I am studying these lessons for a definite purpose.

I am now convinced that I can make myself far more attractive than I am now.

I am going to follow out these instructions from cover to cover.

I am going to put my whole will power into this work.

I am going to accomplish results.

I am convinced that there are no short cuts to the goal of a winning personality, and I must work for it.

I am going to make myself just as attractive as I can in as short a time as possible.

I am going to win the affection and esteem of those I wish.

I am going to make my word as good as my bond.

I am going to keep my eyes open for every one of my defects.

I am going to develop my latent possibilities.

I am going to have entire confidence in myself and in whatever I undertake.

I am going to turn all rebuffs to advantage, by charging them up to Experience and thus profiting by them.

I am convinced that there is no need for failure.

I am confident of my ultimate success.

QUALITIES OF THE MAGNETIC WILL.

Be on the lookout for opportunities to develop your best self at all times.

Don't take risks when you don't have to.

Keep your poise.

Cultivate punctuality.

Do not be satisfied with half-way goals but aim for the highest.

Once every week take a half hour and go to some place where you will not be disturbed and take an inventory of yourself. Note the improvements. Make plans to overcome your failures and plan for bigger successes.

Life is but one unfoldment after another. Our consciousness is ever expanding. Our personal power should be increasing. Everything comes from within. No one can make you magnetic but yourself.

Don't be content to be just an ordinary person and do the things that any one can do. Try to do the things that are beyond the power of the ordinary individual. Develop your intuitive powers. You will never become very magnetic unless you favorably impress and influence the people around you.

You build yourself up by contact with

others. You are quick to notice their good traits and unconsciously you become inspired by them. You associate with a person of strong will, and it stirs your own will power. If your associates have strong confidence, your own confidence becomes strengthened.

There are a very few who do not need a mental stimulant at times. You need to be put in touch with the attractive currents that are always around you, even though unseen. These currents affect your personal magnetism. There is a psychic force, but it is nevertheless a real force that is felt but not seen.

It is this psychic force that is responsible for the fact that we will have confidence in one person and distrust another one.

You will learn all about this psychic force in "Seership: The Science of Knowing the Future."

HOW TO IMPRESS OTHERS

Some persons are like clams. They close up in their shell, and are hard to impress. Sometimes it is really a waste of time to try to influence them. But this is rarely the case, and you can never tell what you can do until you try. You can ultimately reach the most unimpressionable of persons, when you find the key and make an effort. It takes interest to arouse a person. Make them feel that it is really worth their while to achieve what you wish them to do. Make them see what they are and what they might become.

Many a wife has made a great man out of her husband when without her he would have been a very ordinary individual indeed.

Her enthusiasm, and her belief in him, and her mental inspiration have been the invisible forces behind his advancement.

In fact nearly all our great men have admitted that much of their success in life has been due to their wives. Their help was not physical but mental. The mental image of the husband in the successful position which they desired to have them occupy, finally crystallized into fact.

Mind moves the world.

Very often two men will form a partnership, and make a great success, whereas when they worked separately, they were both failures.

The underlying cause of this is due to the fact that they were able to see the weak and the strong points of each other's ability. By cooperation they were enabled to discount the weak points and apply the strong points to the greatest advantage.

That is just what you must do in order to become more magnetic. Be ever ready to discount your weaknesses and plus your strong points.

Before entering into a partnership, whether it is taking a wife, or a business associate, you should study each other carefully.

Find out as much as possible about their past. Find out their personal traits and characteristics. See if they will harmonize with your own. Learn their business capabilities and so forth.

"He who proves indispensable to you as a partner, might be wholly useless or even injurious to another."

General Grant and General Sherman required very different chiefs of staff.

One secret of Napoleon's success may be found in the fact that he was free to make his own appointments. He was thus enabled to select the men whose qualities supplemented his own, and helped him to overcome his own short-comings. Every one is lacking in something.

The genius who can manage everything himself is yet to appear.

You must recognize the qualities in others and attach them to your own power, if you expect to do great things.

The thing that makes for strength in an organization or a partnership is just what makes a complicated machine so effective, namely each distinct and different part of the mechanism fitting into its rightful place and performing its duty without jarring upon the other.

Two persons of strong magnetic power may form an almost invincible partnership by taking as their guide "I am trying to adjust myself to you for what you are at your best."

THE LAW OF DEMAND.

You wish to become magnetic. Then you should demand each and every day that you shall be magnetic. Do this with sincerity and a heartfelt confidence, and in time your demand will be realized.

There is a universal supply from which you may always draw, and by exercising your mental powers you can always attract whatever power you desire. Commit this to memory:

"I demand that measure of magnetism, which my personality and my thought, and my work call for. I am a center. I receive power. I shall gain my goal. I shall make good my demand."

Develop magnetism by demanding it.

SUGGESTIONS IN CHARACTER BUILDING.

Your character is plainly written in your face. "Beautiful thoughts light up the features." What you think has a whole lot to do with how you look. Your appearance has a great deal to do with your attractive power. The physical body is an external expression of the soul. You will find the person with well developed magnetic power, with a good face and a good physical body. Your thoughts and your motives mould your physical appearance.

"As a man thinketh in his heart so is he." The man who lives a pure life will radiate the vibrations of purity and his thoughts will reach those with whom he comes in contact. You can tell a person of pure motives and high ideals by looking at his or her face, better than you can by listening to their words. Your disposition is written plainly on your countenance. A person whose disposition is crabby will have the tell-tale lines on the face; and the mouth will drop down at the corners.

It is not always the handsome person who is the most attractive. A person may have plain features, but by thinking beautiful thoughts he may become very good looking. Few persons understand their possibilities. If you place a limitation upon yourself you will not go beyond that limit. You make yourself what you are. You will never amount to anything if you wait for some one to tell you what to do.

You have been endowed with certain powers but you have to use them if you wish to accomplish your desires. The harder we work the more successful we will be.

HOW TO CONTROL OTHERS.

If you wish to control another, learn his nature and habits and you will then know his weaknesses and disadvantages. Thus you know how to persuade him. The important thing in controlling another, is will power.

Rule I. Belief is the essential thing. In order to throw our whole heart and soul into a project we must believe in it.

Rule II. Confidence is absolutely necessary to success in controlling others.

Rule III. Do not try to control others unless your motives are of the right sort. Remember our warning at the beginning of this book.

Rule IV. It is necessary at times in order to control others, to discover their plans. This you may be able to do, if you have developed your inner faculties. There is a science of knowing the future as demonstrable as any other science. In our work "Seership: The Science of Knowing the Future," we show you how to develop your intuitional nature that you are able to detect the feelings of others; to penetrate their secret motives; and to discover what they try to conceal. Any one may successfully cultivate this ability, if he will study this book.

Rule V. The golden rule must be followed in controlling others. "Do unto others as you would that they should do unto you." If you are polite, cheerful, agreeable, etc., you will be bound to affect those with whom you associate.

Rule VI. A strong will is necessary to control others. You must be able to make them feel as you do; to think as you do, without making it evident that you are suggesting it to them. If you are able to do this, you have a strong magnetic force.

The following suggestions will help you to control others:

1. Don't be jealous or envious of others.

2. If you have an unpleasant opinion of any one keep it to yourself.

3. Don't display your temper.

4. Don't be sarcastic.

5. Don't make remarks about another that you would not make to his face.

6. Don't make remarks about another that will injure him.

7. Do not joke in a way to offend any one.

8. Don't make remarks that you would not make before a lady.

9. Don't make promises unless you feel reasonably certain that you can keep them.

10. Always keep your word, if possible, but if you cannot, don't be afraid to explain why.

11. Don't relate your troubles to others.

12. Remember the chances are that others are not interested in your hobby.

13. Don't try to make a man go against his grain.

14. Let an irritated person return to his normal condition, before contradicting him, even if he is wrong.

15. Don't argue unless you can do so in a peaceful way.

16. Every person is entitled to his own views.

17. Don't sneer at anything.

18. Don't form a hasty opinion.

19. Look for a person's good points, not their bad ones.

20. Always grant a favor if it is right for you to do so.

21. Pleasant words cost you no more effort than unpleasant ones. The former can do you no harm, but the latter can cause you a lot of trouble.

22. There are never two ways equally good. One is always just a little bit the better. Choose the better way.

CHAPTER V.

HOW TO DEVELOP PHYSICAL POWER.

It is just as important for you to develop your body as it is to develop your mind. Most men break down just about the time they begin to know something. If you want to be magnetic you must keep your body in first class shape so that your breath will be sweet, your voice clear and strong, and a little exertion will not fatigue you.

Make it a rule to devote a little time each day to physical development. Not once in a while, but each day. Before going to bed ask yourself these questions:

Have I exercised as I intended?

Have I eaten as I should?

Have I taken the breathing exercises and filled my lungs with fresh air?

Always exercise near an open window. Deep breathing is the very foundation of good health.

Have I had my bath today? Did I bathe every part of my body?

It matters little whether you use cold water or not. The best time for a cold bath is in the morning, and a hot bath at night. Always dry your body thoroughly, and then rub yourself with your hands.

Have I sunned myself?

Have I become excited or hurried myself unnecessarily?

Have I taken sufficient drinking water?

If we would write down these questions in a little book and the answers, for a period of a month we would form the habit and would not have to think of them any more. The subconscious self would see to it that we did it. Life is one big habit.

You will find that all of our really strong magnetic men have taken particular care of the physical body.

The following exercises will take but a few minutes and it will pay you well to take them:

1. Stand erect with chest well up and your hands by your side. Now raise your hands as far as you can over your head, and make your hands meet. Inhale as you raise your hands. Clasp your hands, and hold your

breath as long as you can. Then relax, letting the arms fall gently to the sides, as you exhale.

2. Fold the arms across the chest and raise the body up and down on your toes twelve times but don't let the heels touch the floor. Hold your breath while you are taking the exercises. Aim to raise yourself gently up and down.

3. Stand with your arms stretched out in front with palms together. Swing your arms backward as far as you can, making them come together as near as possible in the back. At the same time rise on your toes and inhale. Each time you come down relax. Put plenty of life into this exercise. This exercise is a great chest developer.

4. Lie down on your back on the floor. Bring the knees up toward the chest as far as you can, inhaling as you do this. Then let your legs go downward until they almost touch the floor, then back again toward the chest. Do this ten times.

5. Stand on the balls of your feet, folding the arms, and sink to a sitting posture. Do this seven times. Inhale as you rise and exhale as you go down.

6. Inhale deeply, fully expanding the chest. While you hold your breath raise your shoulders up and down five times. Then exhale slowly.

7. Stand with hands over head, palms front. Keep the knees stiff and bend forward as far as you can. Try to touch the floor with your fingers. As you go down take a full breath, as you raise up exhale.

8. Now stand and go through the same action, as if you were running. Raise your legs as high as you can. Do this as rapidly as you can.

9. Bend down, touching your fingers to the floor and walk around just as you have seen a dog. This exercise is very good to reduce weight.

These exercises should be taken in the morning the first thing. They will start your blood to circulating and you will have an appetite for your breakfast.

CHAPTER VI.

HOW TO DEVELOP MAGNETISM BY SELF SUGGESTION.

"The habit of expecting great things of ourselves, calls out the best that is in us."

Whatever you undertake to do make up your mind you are going to do it. Never say "I am going to try to do it" but say "I am going to do it." What success would an animal trainer have if he went into a cage of wild beasts shaking with fear, or even with a slight fear, saying to himself "I am going to conquer this beast if I can, but really I don't believe I can do it. It is a pretty tough proposition for a man to attempt to conquer a wild tiger from the jungles of Africa. There may be men who can do it, but I doubt very much whether I can." He would soon be torn to pieces. An animal trainer must have bold courage. He must have a strong eye, because you show it in your eyes if you are afraid. He must have confidence back of that eye because if he betrayed one bit of fear the animal would instantly detect it in his eye and he would probably be instantly killed.

Whatever you undertake go into it with the same kind of confidence the animal trainer has to have.

"A man cannot try with that determination which achieves success unless he actually believes he is going to get what he is working for or an approximate to it."

It is far better not to engage in a thing if you feel in your heart you cannot do it. "It is what we believe we can do that we accomplish or tend to." Make up your mind you are going to make money and you can. But on the other hand if a man starts out without believing he can make money, that one has to be lucky or fall heir to it, he will never make very much money. "Be sure you are right then go ahead" is an old maxim that can always be followed. Let nothing shake your decision but make it a part of your very constitution. It is those with this kind of grit that succeed.

What do the men who achieve great things possess that the ordinary man does not? It is self-confidence. The man who has faith in what he undertakes is the one who succeeds.

We meet persons that we recognize are powerful. It is their faith in themselves that makes us think so. We would not think so well of them if their words were full of doubts and fears. But they radiate power and thus win our confidence, the very first time we see them.

We must not only believe in ourselves but we must have others believe in us. We are dependent upon the belief of others to carry out our plans. We must make them believe in our goods; that we can run a business; pay our bills; and many other things.

This is one of the uses of power—personal magnetism. In this busy age we do not have the time to thoroughly investigate another's ability. If he can make us believe that he can fill the bill, he gets the chance. "The world accepts very largely a man's own estimate of himself until he forfeits its confidence."

Two boys may grow up together. Both have the same opportunities, but one will suddenly branch out and leave the other far behind in the business world. What is the secret? The one had initiative and made opportunities, while the other waited for the opportunity to come to him.

No one will think more of you than you think of yourself. What you are is pictured in your appearance. If you think you are just ordinary you will appear just ordinary. The way you impress yourself is the way you will impress others. You can cultivate any qualities you desire and when you possess them you will express them in your face and manner. You have to feel grand to look your best.

"Confidence is the very basis of all achievement. There is tremendous power in the conviction that we can do a thing." Without confidence there could be no miracles performed.

The Bible says: It was through the faith that Abraham, Moses and all the great characters were able to perform miracles, and do such marvelous things. All through the Bible the importance of faith is emphasized. "According to thy faith be it unto thee."

Faith doubles our powers and multiplies our ability. Without it we can do little. By faith we can come in contact with the Infinite Power and learn the truth from the foundation source.

It has always been the secret of great miracle-workers. Anything that shall increase your confidence in yourself will increase your magnetism.

Most of our men and women who have really amounted to anything have worked along for years and their efforts met with little encouragement, and it seemed as if there were no chance to realize their ambition. But they stuck to their task believing that in some way or other they would succeed. What would an inventor accomplish without faith? Many of them have worked years to perfect something.

Often they have lived upon practically nothing, so as to save the money necessary for experimentation.

If it had not been for persistent effort, they would not have succeeded.

A man came to me who had wealth but no friends. He thought that no one cared for him and they didn't. But there was a reason. He walked about as if he was nobody. He thought he could not do things like other people. He always had an apologetic air and was constantly emphasizing his convictions. He did not expect much of himself and neither did any one else. I took hold of this man and showed him that the good things of this world were intended for him just as much as for any one else. He soon saw that as he began to think more of himself that others began to think more of him, and soon he was a completely changed man. What others think of us has a great deal to do with our place in life.

The larger the faith we have in ourselves the more we will accomplish. We are not limited. Remember:

"There is no inferiority or depravity about the man God made. The only inferiority in us is what we put in ourselves."

Most of us are but a shadow of the man God patterned us after.

We think thoughts that carry us downward instead of upward to the heights of superior realms.

Self depreciation is demoralization. Stand firmly upon your own ground. Don't act as if you were weak-kneed. Assert your divinity and stand erect and look everybody squarely in the face.

The trouble with most persons is that they have fostered their adverse qualities, and have neglected to improve their good qualities.

Don't be one of the "also attended," but from this day on claim your birthright boldly, and have assurance and confidence in yourself.

When you associate with your fellow-men go about as if you amounted to something. Cultivate a strong, vigorous, self-complacent, victorious air. Let other people follow you instead of you following them. There is no reason why you should ever be trailing some one else.

Your employer wants you to stand up like a man, as though you were the equal of any one. He does not like an employee who has an apologetic air. If you have something to say, approach him on the equality of manhood. Those who bow and act as if they were afraid to speak, he does not like.

The "leave-it-all-to-you" employee never amounts to anything. The one who stands up for his rights and shows that he expects to be treated like a man is treated like a man, and forces to the front when an opening comes.

I have taken many a timid, shy, sensitive,

shrinking person, and in a short time made him believe in himself. I just show him that he has all the latent possibilities that any man has. That he can become a man who will be looked up to instead of down upon. I teach him self-faith, and in time all of his other qualities and his strength as well. Keep this ever before your mind. "No man can be greater than his estimate of himself at the moment."

There are men who are real genuises but they never amount to anything because they do not believe they are; a man may have all the ability he needs but if he does not believe that he possesses that ability he will not get results.

Whenever you have a longing to do a certain thing you have the ability to do it. If you will just make up your mind to do it, the achievement will be sure.

What you think you can do, dream you can do, you can do.

Get in the habit of thinking that things are going to turn out right. Never think they are not. This is the secret of successful men. They expect everything they undertake to turn out right, and it does. Despite discouraging indications, they hold tenaciously to their faith and turn defeat into success.

It is a fact that keeping an expectant attitude attracts the thing we long for. This is done in some occult way unknown to any save occultists. Occultists believe that we have guardian angels, who help us when we are worthy of being helped. Whether this is the correct explanation or not, the fact remains that it is true.

What you are looking for is results, and you certainly can get them if you believe you can, and unwaveringly bend your efforts toward that end.

You can wonderfully increase your ability by self-suggestion. There is nothing as harmful as discouraging self-suggestion, such as thinking you are going to fail before you start.

You can control your fate—your destiny. Don't think you were born unlucky or that fate is against you. It is very disastrous to think that way. It is impossible for success to come to you if you think you were born unlucky. If you think of failure and poverty, you are making an impression upon your subconsciousness and you are helping it to develop unfavorable conditions. You are making inevitable that which you should be fighting against. What we call "fate" or "bad luck" is mostly caused by our thought.

On all sides you will see persons succeeding, who have no greater ability than those who are failures. You speak of them as being "lucky, that's all." There is no mysterious power called destiny that helps you unless you do your part. It is you yourself that is keeping you back.

You must have the right mental attitude. You must demand your right. We are all superb beings with infinite divine possibilities. Your Creator is responsible for you. You have within you omnipotent possibilities. You have inherited divine qualities but they need to be developed.

THE SECRET OF THE LAW OF FINANCIAL ATTRACTION.

It is possible for you to attract Money, Love, Business, Health, Wisdom, Harmony, Happiness, Strength, and anything else you wish. You have everything within you. You are a vibrating center of magnetic force and vitality. By your desire, belief, and will, you attract the elements of all things that affect you.

There is a tremendous, creative, producing power in the perpetual rousing of the mind along the line of the desire. It develops a marvelous power to attract the things you want. You can realize your highest desires now.

To do this you must become a conscious center of veritable force. Through vibration you can change every undesirable thing of your environment and remove your weakness. Your thoughts are powerfully creative, and by constant thinking you add a greater potency to the words you speak.

Science tells us that no vibration ever ceases, but the influence goes on forever. By attractive vibratory powers you can improve your condition.

What ever you wish, think of that and never swerve from your desire. If you want to win the love of some one, never allow yourself to think you can not. If you want prosperity, think prosperity, and never of poverty. Act as if you were already prosperous, dress as though you were. Make a mental picture of what you want to be and then see it develop.

By all means be brave, courageous, fearless. Be afraid of nothing. If you are naturally timid and shy, suggest to yourself that there is no reason why you should be afraid of anything or anybody. Hold up your head and rouse your real manhood or womanhood. Throw back your shoulders. Assert your birthright of power and equality.

Look at yourself in the glass. If you think you have a weak face you have one because you think it is weak. Think it is strong and you can make it strong. By a little daily practice you can cultivate courage and self-confidence and in a short time you can build up your timid character into a strong, bold one.

The more confidence you have in yourself, the more ability you will develop. Do everything you can to increase your self-confidence. The power of suggestion will help you a great deal. Never pay any attention to adverse suggestions of others. You know yourself better than they do. There is a lot of power in personal suggestion. Always act in such a way that others will see your improvement. It is a wonderful help to get to be considered progressive and a person who can accomplish things.

When you meet a friend on the street he makes a mental picture of how you appeared to him. If you seem to belong to the Booster's club he is quick to notice it. If you appear to be a member of the Down-and-Outs he is equally observant. His opinion of you he reflects back to you. If this is as it should be, you receive his suggestion and it will help you. Show by your manner that you are climbing all the time, a little bit higher—that you are a coming man. Never for a moment think poorly of yourself, or any one else.

There is a place for you in the world, and fill it like a man. *You* can do the great things in life just as well as any one else. By persistently holding positive thoughts, you can create whatever you wish. Keep in mind: "Nothing comes without a sufficient cause and that cause is mental."

Our character is continually being molded by our thought.

CHAPTER VII.

HOW TO USE YOUR PERSONALITY TO WIN THE AFFECTION OF THE OPPOSITE SEX.

There are, of course, numerous ways of gaining the admiration of those you wish, but I will give only the most practical ones. First we should study the character of the one we wish to win. Love is first started into flame by sympathy—or by liking or pretending to like the things the other one likes. But right here let me warn you—never pretend. Oftimes to be sure, affection has been won by pretending to be in complete sympathy with the one desired, only to prove disastrous later on.

Pretensions may do during a short courtship, but they will not make good after marriage. This is the cause of much of the unhappiness in married life. In true love there is no pretense. There is entire sympathy with each other's aims and ideals. It is not meant that both should be the same in character; neither is it expected that their wishes shall be identical in all respects; but there must be a general inclination toward the same desires and tastes.

THE DEVELOPMENT OF YOUR FASCINATING QUALITIES.

The development of your powers of fascination and how to use them to the best advantage is a study in itself. It is the inherent desire of every natural man to win the affections of a good woman, as it is the desire of every woman to win some good man. Until they do, they feel that they have not enjoyed real happiness. I will mention some of the qualities women admire in a man.

In the first place, he must be a man through and through. He must treat women in a manly way, and fully respect them. He must be careful in his language, never to say anything that he would not say in the presence of his mother.

The kind of woman a man admires is a womanly woman. She must at all times keep her dignity. Be a lady at all times. A man may sometimes enjoy the companionship of a "good fellow" sort of girl, but she is not the kind he chooses for a wife. He selects the one he holds in the very highest esteem and in whom he has confidence; the one who conducts herself with modesty and good sense.

A man is never attracted to the cold appearing girl with an exaggerated dignity and an unsociable manner; nor the girl who always seems to be afraid of her reputation, as though it were something too frail to withstand everyday use; nor the one who seems always suspicious of actions and words.

Real love can only be experienced by the manly man and the womanly woman. They are the only ones who have the capabilities of the real lover. Mere attraction between male and female is not real love, although it may develop into love by the great law.

There are very few men and women who know anything of the real qualities they should possess. They do not know how to select one of the opposite sex worthy of real love. But any one who reads this work cannot give this lack of knowledge as an excuse hereafter. Real love makes a wonderful change in a person. A couple meet and in time they become very much fascinated with each other. The man obtains power over the girl. She is willing to leave what has been the dearest things in her life—her home and family—for the man she loves. She will sacrifice everything for him. The same is true of the man. It is in every one's nature to want a home of his own; a fireside to preside over. It is here that a woman develops her real self.

We will now study the man as he really is, with all his weaknesses.

Most men are fond of good meals. A wife can find out what he likes, and if she uses tact she can always have something that he likes. Probably the biggest mistake a woman can make is to keep herself untidy. You hear a woman say, "I wouldn't give the snap of my finger for a man who wouldn't love me the same in my old clothes as he would in my best clothes." They don't stop to think that there is a very big difference between a fair face free from scowls, hair nicely arranged, an attractive dress, a pleasant smile, which is what the man sees in his courting days, and the way she usually keeps herself after marriage.

You will find that the wife who holds her

husband's affections is the one who takes the same pride in her appearance after marriage as before, and who retains her modesty.

The former is what he thinks he is getting, but when he finds that she is the reverse, his affection is likely to change.

It would be well if all men could see their future wives in their home when they are not expecting any one. If a girl is in the habit of keeping herself untidy in her parents' home, she will in her own home when she has one.

It is easy enough to win a husband or a wife, but there are few who are capable of keeping them feeling that they are the one person in the world that they want, as in the days of courtship. Every man and woman possesses good qualities and the secret of a happy home is to draw out these good qualities. Far more may be accomplished with gentleness than with brute force.

There is a certain knack of doing things that brings results. The person who uses tact may get his own way about things and yet make it appear that it is the other one who is leading. Headstrong persons have to be managed in this way.

There is a wonderful magic in love. It brings the only real happiness in the world. Marriage should bring happiness, and it would bring it if each one would develop the qualities of true manhood and true womanhood. Every man should have a kind and devoted wife who believes in him with an unswerving confidence. Every woman should have a real man for a husband, one who is worthy of her respect and affection.

People as a rule are not careful enough of their associates. "A man is known by the company he keeps;" so is a woman. We should study new acquaintances carefully and discover whether or not their friendship would be valuable. It is not always those with the highest education who make the best men and women. It is those with the greatest amount of common sense who are the most desirable. With careful observation you can soon decide whether a person has desirable traits of character. At first a person may pretend to be what he is not, but gradually the real nature is disclosed, and especially if he be permitted to do the talking. One thing will disclose another until finally the truth is revealed.

The young girl who is receiving attention from a young man should permit no liberties from the very beginning. There is an old saying to the effect that certain things should be "nipped in the bud." Keep the bud from becoming a blossom.

"Coming events cast their shadows before," and a person's character and purpose may generally be read at a glance. Men and women of today are more careful in selecting partners for marriage, and still divorces are becoming more numerous. That wonderful love that sways men and takes them "off their feet" must be better controlled.

CHAPTER VIII.

LOVE AND COURTSHIP.

Love is the foundation of all happiness. It is the regenerating process of nature. Marriage is one of the most important steps in life which a man or a woman can take. It can be the blessing of both, or it can be the ruination of both. It is a deplorable fact that observation reveals more unhappy marriages than happy ones. There must be a cause for this, and there is. The fault lies in the fact that couples marry when they are not suited to each other. If they would be more careful in their selections and instead of marrying too early in life, take time to improve themselves, as suggested in this work, there would be more happy marriages. I trust that all young men and women who read this work will take advantage of the instruction before marrying.

If one should believe in these teachings

and the proposed partner should not, and refuse to develop along these lines, they are not suited to each other. No one should marry a man or woman just because they are engaged to do so. Unless you are thoroughly satisfied that your choice is a good one, don't marry. The time to reflect is before you are engaged, but if you have changed your mind up to the wedding day, you should not enter into the marriage.

Marriage is a serious proposition and deserves careful study. Your own happiness and the happiness of others depends upon it. The vast majority enter into marriage without knowing or thinking of the responsibilities. They take a blind chance, and the result is that they "get stung."

Nature's plan was no doubt that every one should marry. The majority of persons want to marry, but there are thousands and hundreds of thousands who do not marry, and as a result live a lonely and miserable existence.

There must be a cause for this and it may generally be found in the fact that so many lack the power of attraction, or personal magnetism. You will never find the magnetic person devoid of plenty of opportunities to marry. They are able to fascinate and control the minds of others.

THE CONTROL OF ONE MIND OVER ANOTHER.

It seems a remarkable thing that some persons are able to control others, but they do not know they can, and those whom they control do not realize that they are being controlled.

You can be almost completely under the control of another, acting just as he or she desires you to act, and at the same time you think you are obeying your own will.

SOME SECRETS OF INFLUENCING OTHERS.

You may completely control the actions of others without their suspecting it. They may think all the time that they are doing just what they want to do, instead of which they are carrying out your wishes.

A person may be so completely under the influence of another that his body is controlled by that other's thought. This is done when the hypnotist puts a subject to sleep. While under this influence the subject's mind is put out of use. The operator controls the body of his subject as if it were his own. The only way to make sure that you are not being controlled by another person is to form the habit of sitting alone and meditating before you decide any great problem. The habit that some men have of taking a proposition under advisement over night is a good one. You will find the answer you have arrived at by morning will be the best one for your guidance. The question often arises, "How is it that each year finds fewer and fewer powerful men are being developed?" Or why is it that there are not more magnetic men born? We cannot speak of the past, but I can speak of the present. I think that men of today as a general thing have more magnetism than ever before. This is the natural process of evolution. As man develops higher and higher in the scale of progression, he has more magnetism. The men and women in the lower scale of development have little magnetism. But still you say that this does not answer your question, so I will explain more specifically.

In the present age we have to go at such a pace that after a day's work our magnetic force is greatly diminished. Instead of retiring early and spending the night in quiet repose, giving nature a chance to restore this magnetism, more work is done at night. The business is carried home and the work for the next day is planned. If not business, then a round of exciting festivities is indulged in. The result is that instead of restoring the lost magnetism, more is expended. This is why people get old before they should. They do not give nature time to restore their forces.

The Bible tells us that after Jesus had administered to a great many of the sick, he was compelled to withdraw into the mountains or the country until his supply of magnetic force was replenished. The person who does his work in the office and forgets it when he leaves his place of business will accomplish a great deal more in the long run, because he will be able to put more real energy into his work.

The evenings should be spent in some phase of enjoyment very different from the work of the day. To do your best work you must enjoy yourself. Enjoyment means interest in life. Those who do not enjoy life do not really live, and lead an uninteresting life. When a man works until he is all tired out it takes him a longer time to recuperate. Few of us realize how important a part in our life the mind takes. If you have worked hard all day and are tired from your efforts, and then spend the evening with others who are equally de-magnetized, talking of their work and expressing their worn-out feelings, you will feel worse than ever. But, on the other hand, no matter how tired you are, if you go home and take a bath and change your clothes, and spend the evening in company with others who have done likewise, you will find that you not only enjoy associating with them, but your magnetism is restored. There is an exchange of magnetic force that is mutually helpful. Wives make a big mistake when they work all day and then show that they are tired out at night. They might have worked much less hard if they had but thought they could. Don't greet your husband with that tired expression. He is already tired, and when he takes on your mental condition his weariness is that much exaggerated.

When a person is over-tired the least little thing makes him irritable and cross. Those of you who are married or going to be, think this over. It will make a big difference in your life if you heed this warning.

Whatever you wish for, you must place yourself in a receptive state to receive. If you wish more energy, then be determined that you are going to have more, and not think you are going to have less.

There is a law that will bring to you whatever you wish. But this law can also work in accordance with the principle of opposites and take from you what you want. The secret is in learning how to control the elements of success.

One should remember that the first law of nature is self-preservation. If this law is heeded you will be able to help others also. Every one should try to live his own life and not burden others with his troubles or be burdened by the troubles of others. It does them no good, and it does you harm.

Descartes says that "if all the troubles in the world were laid in a heap, each person would take his own as preferable." This is due to the fact that each one's problem is for his own solution. This does not mean that we are to be cold hearted and unsympathetic, but merely that mutual help does not necessarily involve the need of being burdened with the troubles of those about us.

It seems a curious fact that those who possess the greatest amount of magnetic development are not the ones who are most intellectually developed. There are men with little education who can hold immense audiences spellbound. Their force is so strong that people give no thought to the grammatical errors they make. The spirit of their utterances holds their auditors in rapt attention. The late Dwight L. Moody is an illustrious example of this power of personal magnetism. Unaided by education or the gift of oratorical phraseology, Moody moved vast congregations to feel the power of his presence.

On the other hand, there are men who through much learning have so perfected themselves that they speak with distinction and logic and yet they cannot hold their audiences like the other type of men.

This is the difference between a magnetic man and one who is not. Magnetism plays a bigger part in your life than you have any idea of.

Some day magnetism will be developed as it should be. This invisible force is found and used everywhere, consciously and unconsciously. In business it plays a wonderful part. It is used in curing disease. In fact its uses are unlimited. It is only necessary to bring out your latent abilities.

CHAPTER IX.

WHAT CONSTITUTES A PLEASING PERSONALITY.

If you want to be noticed you must present a good appearance and a pleasing address and manner. It matters not how much money a person has, or how high his social standing, or how long his family tree. If he does not present a pleasing personality he will never be a favorite with either sex. You have no idea how important a part personal appearance will play in your future.

Personality is a great factor in business success. One of our leading merchants has stated: "The man or woman wishing to present to me a business proposition must have a good address, an agreeable manner and pleasing appearance, or he will not get a hearing, no matter how good the proposition may be. The reason for this is a simple and natural one. It would be impossible to give a hearing to even half the number of persons who approach a business man with schemes. Therefore, as I must reject the majority of propositions offered me, I reject without a hearing all those that are not presented by persons who have an agreeable manner and good address. I take it for granted that a first class proposition will be presented by a first class man, and vice versa.''

There is no doubt that this man loses some good opportunities, as there are many men who have worked out good propositions who have failed to realize the importance of developing a pleasing personality, but the opinion voiced by the man here quoted, is the way of the world.

The development of a pleasing personality is as important as any of the studies taught in our universities. In fact, personality should be taught in our schools and universities.

The remaining part of this chapter will be devoted to showing you how to improve your personality.

We all know that a good appearance goes a long way with almost every one. A position is advertised. The employer comes out and selects different ones for an interview. Does he select those who are slovenly in their dress? No. He selects those who are careful of their appearance. The person who gets a position and the one who gets promoted is the one who shows by his appearance and manner that he possesses self-respect.

An employer sizing up applicants for positions first considers the following:

Neatness of dress, the manner, cleanliness of person, and, lastly, references. If the applicant does not have the former qualifications he takes no time to investigate references.

No matter how much natural ability you may have, you will not get the opportunity to show this if you are not careful of your appearance, so as to make a good impression.

The value of a first impression cannot be overestimated. It is a great deal easier to make a favorable first impression than to overcome an unfavorable one.

You are studying this work to develop magnetism, and it has been my aim to touch upon everything that can in any way help you. It may be your present opinion that every one tries to keep their personal appearance the best they can. But keep your eyes open for a few days and see. Watch yourself and see if you should not wear cleaner linen; if your shoes could not look better; if your clothes are pressed and free from spots; if your hair is kept brushed and combed; if your fingernails are perfectly clean. All these things must be given attention. People forget that they are judged by their appearance. A person's personality shows in everything he does. There is no exception to this rule. The world judges a man's merits very largely by his appearance, and the sooner you learn this the better.

In America and England and in all progressive countries personal appearance plays an important part. The following is reprinted from the London Drapers' Record: "Wherever a marked personal care is exhibited for the cleanliness of the person and neatness in dress, there is also most always found extra carefulness as regards the finish of the work done. Work people whose personal habits are slovenly produce slovenly work; those who are careful of their appearance are equally careful of the work they turn out. And probably what is true of the workman is equally true of the region behind the counter.

Is it not a fact that a smart salesman is usually rather particular about his dress? Is averse to wearing dingy collars, frayed cuffs and faded ties? The truth of the matter seems to be that extra care as regards personal habits and general appearance is as a rule indicative of a certain alertness of mind, which shows itself antagonistic to slovenliness of all kinds."

There are many people who have failed in life because they did not realize the importance of keeping up a good appearance. If you want to climb the ladder of success you must be able to make a good impression upon those with whom you come in contact. Many are traveling on the downward path because of carelessness of their appearance or a disagreeable manner. People condemn themselves before they say a word. They are not given a chance to display their ability. Very few persons are able to judge a diamond in the rough. They are looking for the finished product. In the long run you will not be given a hearing if your appearance is not good. If you want to succeed you must be careful of the smallest detail of dress; you must improve your manner and appearance in general.

You will find very few who possess all the qualities that go to make up a good appearance. The vast majority will lack some of the essentials. Some will be well dressed, others well mannered, others cheerful, others well groomed, but you will seldom find one with all these. You must have them all to make a pleasing appearance. If you lack one of them you will not have an attractive personality.

There is really no good reason why the average person should not have all these essential qualities. You may think it takes a lot of money to be well dressed. But after careful study I have come to the conclusion that it costs just about as much to be badly dressed as to be well dressed.

Most of the people who will read this work have enough clothes to look well. I have seen people wearing clothes that cost a great deal of money, yet they did not look as well as those with cheaper clothes. It is not so much what we spend for clothes as it is taste and good judgment that makes the man or woman look well dressed. The following illustration will show that it does not require so much money as care in order to make a good appearance:

"A young man bearing a letter of introduction and recommendation called at an office seeking a position as bookkeeper. The letter was invalidated by the young man's appearance. Though but twenty-five, his shoulders were bent, his sentences were uncertain, his eyes wavering, his linens soiled, his necktie askew, his teeth disgustingly black, his face unshaven, his fingernails dirty and his clothes unbrushed. A business man would not have such a frowsy man about his office. Every one of these faults could have been corrected without expense other than care. Poverty could not be pleaded as excuse. That young man will have a hard time all his life, and probably blame his friends, the times, and his luck for his failure, when his disgusting slovenliness is responsible."

One of the great drawbacks to young men seeking advancement in the business world is their carelessness in regard to dress and cleanliness and their neglect to develop an attractive personality.

We hear people say that little things should not influence a person of good judgment.

True they should not, but they do. It is the little things that make up the larger ones. There are some who can see the real man despite his defects, but they are few. The average man and woman judges by external appearance.

It is useless to say we should not judge a man by his outward appearance. The fact remains that we do. The only thing to do is to pay attention to your outward appearance and improve it all you can. Much of your success depends on your appearance. The magnetic man will always be well-groomed. He does not leave his room in the morning until his toilet is complete and irreproachable. By paying careful attention to your appearance for a short time you will soon get this so fixed in your mind that you will find it is just as easy to keep yourself looking well as not to do so.

The successful corporations are paying more attention to the appearance of their employees than ever before. They know it does not argue well for their business to have slouchy looking people in their office, and then, again, as has been previously stated, they know that people who are slouchy in appearance will turn out slouchy work. Their habits make their character. Employers want employees who take pride in themselves and have enough self-respect to take proper care of their teeth and hands; who never miss a daily bath, so that the portion of their body that is not exposed will be just as clean as that which is.

"Cleanliness of soul and body raises man to his highest estate."

After you once realize the important part your appearance makes in your life, you will surely keep yourself clean and make yourself as attractive as possible. This is not a book on health, but I feel it is imperative that I should call your attention to the following:

Rule 1. Take a daily bath. This will make the skin clean and wholesome and without which health is impossible.

Rule 2. Brush the teeth every morning and night; keeping the teeth in good condition is a very simple matter.

Rule 3. The hair should be brushed every day. If the hair is very oily it should be washed every two weeks with some good soap. It will help if you add a little ammonia to the water.

Rule 4. If you can afford it, have your fingernails manicured once every two weeks. If you cannot afford this, do it yourself. You can buy a manicuring set very cheap.

All these things you, of course, know you should do, but I am reminding you to see that you do it. If you are careless of your teeth your breath soon becomes foul. Under no circumstances neglect your teeth. It is easy to preserve them. It is, of course, best to brush them after each meal, so as to remove all particles of food. It is best to use tepid water, and the best, and also the cheapest, thing to use is powdered orris root, which will also help to keep the breath sweet. It is a good thing to occasionally clean the teeth with fine salt.

WHY DRESS IS SO IMPORTANT.

For the apparel oft proclaims the man.—Shakespeare.

We are all charmed by the neatness of the person.—Ovid.

"Neither virtue nor ability will make you appear like a gentleman if your dress is slovenly and improper."

The one who does not pay attention to his

personal appearance is making a fatal error. It is either his recommendation or it is the reverse. The more prosperous you look the more success you will attract. Let the salesman or professional man wear out-of-date or shabby clothes and he will find he will not do near as well as if he was well dressed. He does not have confidence in himself. His magnetism is not as strong because he does not feel at his best. Thomas B. Bryan says: "Adequate and becoming apparel makes a stronger impression on the person wearing the clothes than on any who observes it. If every business man now going about his affairs in garments which are a little below the reasonable standard of presentableness could be clothed with those which fully meet the requirement, the business world would feel a sudden and unaccountable pulse of no mean proportion." Clothes don't make the man, but they do help him make an impression. When I speak so much of clothes I do not mean that you should try to be a Beau Brummel. To spend money foolishly on dress is a mistake, but to spend enough money so that you will look well is something every one should do.

THE VALUE OF POLITENESS.

Edward C. Simmons, a self-made millionaire, says: "Politeness is the cheapest and often the best capital in the world. You can draw upon it in unlimited quantities. It pays a large interest and costs nothing. I would say to all young men: 'Do not keep your politeness down in your boots where no one can see it. Let it out. Keep it in front of you and all around you. If you are polite to everybody, everybody will be polite to you. * * *' Politeness will also contribute much to your success in commercial life."

The man or woman who wants to be popular must be polite. "Politeness," says the editor of a leading New York newspaper, "is really a national asset. Polite people are a cheerful people, and cheerfulness means good health. Nothing interferes so much with a man's physical well-being as constant anger and irritation. And nothing facilitates the transaction of business so much as widespread politeness. We suggest to hundreds of thousands of active young and middle-aged men who rush about the streets, jostling each other, often quarreling, nearly always in antagonistic spirit, that they give a trial to actual politeness. The next time a man runs into you or steps on your feet, or appears to crowd you when you are probably crowding him, just try an experiment. Instead of rebuking him in some slangy way, express your regret at having crowded him, and assume that the fault is yours. You will be surprised to see the expression of his face change and to learn that the habit of politeness is most catching."

As a rule it is just as easy to please as not to. Those whom you do please will be your boosters and you will never know when you may need a boost.

It never pays to be uncivil to any one. You can never tell when you may want the help of the person to whom you were uncivil. Incivility can do you no good, if it does you no harm. You will find this a mighty good philosophy: I will never do anything that cannot do me any good but may do me harm. We often hear the expression, "He is a little queer, but in his heart he is all right." Now, what is the use of your being all right if you do not appear to be?

A rude manner quickly creates prejudice

and closes hearts and doors against you. Nothing that I can think of is a greater stumbling block to the development of personal magnetism. There are very few who cannot improve their manners. If you think you act all right, just keep track of yourself for a few days. Lose all that roughness the minute something goes wrong. Instead of getting angry, smile. Let the other fellow get angry and notice how much better you look than he.

Little things have been the cause of murders, of wars. When you are angry you lose your reasoning power. Trifles loom large. Whenever you see a person with a gracious personality watch him closely. You can learn much from him. Also watch those with a gruff, uncouth bearing. You may be guilty of some of the same things that he is doing, only in a lesser degree.

A little grace of manner, a little refinement, a little more self-control would make a big difference in the majority of the people. Good manners play such an important part in our lives that it seems a pity that its importance is not more fully recognized and taught in our schools and colleges; also that parents do not set their children the example and teach it in the home.

A person may be very highly educated, but if he is rude and boorish he will not make a very good impression on those he meets.

A boy may spend a great many years in school and college and yet never be taught the advantages of cultivating a pleasing personality. It is very easy to acquire the habit of being polite, kind and gracious to every one. It will help you a great deal and cost you nothing. It will make you popular and bring you happiness. The person with a charming personality is more irresistible than the one with a good looking face. He is able to accomplish the seemingly impossible. Personality has conquered hard-headed men. It has won great statesmen. It has influenced the destiny of nations. It has come out victorious where force, money, beauty, knowledge and wealth have failed.

QUALITIES OF A GENTLEMAN.

When Thomas F. Bayard retired from office, the London Spectator said: "What the English people admired most in the retiring minister were his qualities as a gentleman. It was these qualities, combined with his integrity and general ability, which made him so interesting a figure in our nation's life. He had the personal distinction, the dignity of carriage, the devotion to duty, the charm of manner, which are the finest qualities of a gentleman.''

A man may have a bad reputation; you may think his views are all wrong, his principles bad, but while you are in his presence his eloquent manner will make you think he cannot be as bad as you have heard him pictured. Many people went to hear the late Bob Ingersoll speak who did not believe in his philosophy, who were utterly opposed to everything he said, yet in spite of this they went to him, paid their good money to see him, applauded him and came out thinking that perhaps he might be right in some things after all. Perhaps when they went home they reverted to their former opinion of him. They had been conquered by his matchless oratory and exquisite charm of manner.

The old saying will always be true, ''Our manners give their whole form and color to our lives,'' and that ''according to their qualities they aid morals, they supply them, or they totally destroy them."

Refinement, grace and charm are the weapons of personal magnetism. They can make or mar you. In this day of hustle, bustle, and scurry, less attention is being paid to manners than ever before. It is a case of "get out of my way." People go about with long faces. You don't see them smile. At the slightest provocation they lose their self-control and give utterance to some hasty remark. The golden rule is practiced very little. But if it were practiced universally, how different life would be. We would be enriched and uplifted, made cheerful and happy, and it would cost us nothing but a little effort.

Before closing this chapter I want again to advise you to cultivate good manners, and you will have friends and popularity. You will be happy and make others happy. "The art of pleasing is the art of rising in the world."

We judge a person as we see him. He is his own walking advertisement. As we study him every word he speaks, everything he wears, everything he does, every movement of his body is a key to himself. By studying these things in people we learn what kind of men and women they are.

We impress people differently. Some will be favorably impressed by some trait which will jar upon others. To become magnetic and popular we must be careful that we do not have habits that displease. You want nothing to impede your progress. A person has to be careful not only of what he speaks, but how he speaks. "There is a man I know," says a friend, "who employs a great number of persons in the course of a year, and yet never sees the face of one of them. He sits behind curtains in his office and listens to the voice of the applicant for a position as he responds to the questions put by his representative. 'I believe in the human voice,' he says; 'it doesn't lie, as does the manner or the general facial expression. I do not care what a man says. Indeed, I never listen to his words. What I want to hear is the sound of his voice—its intonation, its pitch. You can conceal your real character for a time by your actions. But in your voice God has written your true character infallibly. It never has betrayed me.' "

There is no question that we are wonderfully influenced by the quality of the voice. In the future more attention will be paid to the voice than has been in the past. We simply cannot like a person with a voice that rasps our nerves.

Of all our physical qualities the most important is the eye. The eye is the mirror of the soul, and through them our character is disclosed. If a person cannot look us in the eye we cannot have any confidence in him.

One of our large employers of men and women says he is guided in his selection of applicants for positions largely by the expression of their eyes. "There," he says, "I read honesty or deceit, intelligence or dullness, courage or cowardice. I place little confidence in a shifty-eyed individual, even though he have every other point in his favor. A direct glance from clear bright eyes wins and compels respect. Clear, honest eyes, indicating a sound and vigorous mind and body, are most desirable in salespeople; they will attract customers."

It is an old saying that the eye cannot lie. A criminal is not afraid that he will tell of his deed, but he is afraid that he will display his guilt in his eye. That is the reason he will not look at you. Police officers understand this. That is why they get squarely in front of a criminal and compel him to look them in the eye. When they are innocent they prove their innocence right there. By this test many an officer has reported to his chief that his suspect was not guilty and should be freed.

You may teach your faculties to lie, your manner to deceive, but it is hard to make the eye tell anything but the truth. The eye needs training as well as every other part of the body. Your personality will be greatly improved by an honest, straight-forward look. No one can hurt you with their eyes. When you are talking to a person look him straight in the eye.

OTHER REQUIREMENTS FOR A PLEASING PERSONALITY.

We see people whom you immediately say are "stuck on themselves." A person of this kind is never well liked. One of the first requirements for a pleasing personality is to appear unconscious that you have a pleasing personality.

As you walk along, walk with an air of determination. Never appear shy. If you see some one you are acquainted with, don't be afraid to speak first. You may meet a person and think that he is looking right at you and yet he does not offer to speak. The chances are he did not see you. If he did see you he may have been so absorbed in thought that, although he looked you straight in the eye, he was altogether unconscious of the fact. This is quite possible. Don't wait for the other to speak, but speak yourself. Never display any shyness. Never walk along the street with your eyes on the ground or look the other way when you see some one coming whom you know. The chances are he will think that you saw him and turned your eyes away on purpose. This will put you in an unfavorable light.

All timidity should be overcome. The timid person cannot appear natural. He will appear stiff, cold, reserved, and yet he may be just the opposite. A person who is embarrassed will not display his originality. He is afraid he will attract attention. People of this kind will never attract admiration. Their timidity closes the doors against admiration. They continually put off what should be done at once. In this way they lose many opportunities they might have grasped. Shyness has been the cause of defeat for many otherwise brilliant men and women.

Shy people never have much magnetism because of their retiring nature. They repel friends instead of attracting them. They are afraid to act as they feel. The result is that they are not sought after, no matter how fine a character they may be. There are strong men with remarkable ability that, no matter what the gathering, they soon find themselves all alone. They do not wish to be thus neglected, but they involuntarily choose their conditions. These same men have the qualities that would make them popular, but they are not developed. There are others who have not half their natural qualifications who far outstrip them in both business and social life.

Shyness is a purely mental characteristic. Most of it is caused by imagination. When a person is shy he thinks that whatever he does or says some one is looking at him or listening to him. I have cured a number of my students by a very simple method. It is this: No matter what you are doing, never look around to see if some one is watching you. Whatever you have to do go ahead and do it and never think whether you are being watched or not. I generally require a student to do something in his own way, and have several people watch him and making remarks about his awkwardness. I then show that his awkwardness is due to the fact that he is watching the other people instead of paying attention to what he is doing. In a very short time he loses that awkwardness entirely. Instead of avoiding people, make it a habit to always have something to say. When you go to a party or reception, stay in the "bunch" and be one of them. Be ready to do what the rest of the company want to do. "I was very shy," said Sydney Smith, "but it was not long before I made two very useful discoveries. First, that all mankind was not solely employed in watching me, and that shamming was of no use; that the world was very clear sighted, and soon estimated a man at his true value. This cured me."

You cannot take your rightful place in life until you have conquered your shyness. You will find it a little difficult at first, but every time you conquer yourself when you notice a tendency to shrink from something you should do, you will find it easier the next time. Soon you will see that there is no reason why you should be shy and boldness will accomplish a great deal more. The following will help you if you will commit it to memory:

I am a free and powerful child of God. I am a prince or princess in my right, heir to the king of kings. I am the equal of any one.

You will then never feel inferior to any one. You will never fear that they will criticize or make fun of you. Don't be afraid of public opinion or what others might say or think about you.

I had for a student a man who was compelled to be before the public a great deal of the time. He was naturally very sensitive and suffered a great deal over what others might say about him. He had tried various methods to overcome this weakness, but could not. Finally, after being almost driven to desperation, he came to me. He said: "Professor Dumont, I want you to cure me of my awkwardness and teach me how to act, so that I will not be ridiculed and criticized all the time. It seems no matter what I do I blunder." I told him that would be a very easy thing to do. My first instruction was to become utterly indifferent to people's opinion of him; to ignore what they said or thought about him. After the first lesson his shyness began to disappear and in a short time he was entirely free from it. He acted in his natural way, he did what he thought was right instead of wondering what people would think or say about him.

A young, unhappy, awkward, self-conscious, blundering girl was transformed into a graceful, attractive, well-poised girl who could meet strangers with perfect composure and carry herself with ease and grace in any society. I first told her her possibilities, and then aroused her to conquor the weakness which had been the cause of her unhappiness. I then had her repeat the following at least ten times a day:

I am going to meet people as if they were human beings, not judgment seats. I will not always wonder whether I shall please them. I shall wonder a little just at first whether they are going to please me.

This will help any one suffering from self-consciousness. The one great secret of curing shyness is to get your mind away from yourself. Don't ever think about the impression you may be making on someone. Just be your own natural self, and you will be found more attractive. Always aim to be original. Originality makes the dullest person interesting. Some points to be remembered:

Form the habit of saying pleasant things to others.

Don't look for a person's faults; look for their good traits.

If someone has done you a wrong, forget it. Always be ready to forgive.

You can never be magnetic while you hold a grudge or cherish an unforgiving spirit.

You will never be popular if you look for the bad instead of the good in others.

Aim to be good-natured and cheerful.

Be every ready with a smile and a good word for those who need sympathy.

CHAPTER X.

THE WONDERFUL POWER WITHIN US.

You can do everything you ever dreamed or imagined you could do if you could utilize all the power within you.

This may sound impossible to some. I was at a theater where a hypnotist was performing when I first heard the above thought expressed. The hypnotist saw a man that showed he doubted him. The hypnotist said, "My friend, I see you doubt what I have just said." "I do," said the man. "My friend, I am going to prove that I can do the impossible. You see that young man sitting beside you? Do you think it would be possible for me to take him and put his head on one chair and his feet on another and then have five men stand on his stomach?" "No, sir, that would be impossible, as he is nothing but a frail youth." The hypnotist said, "All right, you consider this impossible, do you not?" "Yes, sir," was the answer.

The hypnotist asked the young man to step up on the stage. He hypnotized him, placed his head on one chair, his feet on another. He asked six good sized men from the audience to stand on the man, which they did without any harm to the young man.

This only shows what would be possible for you to do if you could just think you could do it. While under the influence of hypnotism the subject loses control of his mental powers. The hypnotist gains control and can make the subject do anything that he makes a strong mental picture of.

If this young man could support six men while under the influence of hypnotism, he could do it in his natural state if he could just think he could do it.

You might search the whole world and could not find a strong man that could equal this feat of the youth. Where did the power come from? The hypnotist could not have conveyed this power to him. He must have had it in him. This was just what he had. The hypnotist just made him use it.

This example is given to show you that you

have wonderful latent powers and if you can only learn to use them you will be able to do some wonderful things.

We all have a great wealth of power within us. It only takes some emergency to bring it out. There have been cases where invalids of years have gotten up and walked out when the house caught on fire and they knew there was no one near to help them. They have escaped in such a way that it would seem only the strongest of men could have done. How could they suddenly possess this power? It shows that we have wonderful power within us if we could only command it.

You have powers within, which if you could discover and use would make of you everything you ever hoped or desired to be.

THE CONTROL OF OUR MENTAL FORCES.

In time to come we will have fewer physical doctors and more mental doctors.

We know that by driving away pessimistic, bitter and angry thoughts, we can avoid sickness and misfortune to a great extent. We will realize that by cultivating a kinder and happier frame of mind it will bring us health and good luck.

The time will come when a mental doctor will be able to analyze a person's character and tell what thoughts are discordant and vicious and injuring his personality.

We hear people talking of their fate. The weak ones are controlled by the forces. The strong ones control the forces. Things have just the power over us that we permit them.

When you are unhappy, distressed, blue and worried, it is a mental poison only. You should be able to relieve it by mental processes.

In former ages we were not advanced far enough to understand the philosophy of loving our enemies, but now we can understand it. If we hate them we are the sufferers. If we love them we are the beneficiaries.

By changing hatred thoughts to love we make a friend instead of an enemy. Love cannot make an enemy. An enemy can do us a lot of harm. We are the gainers you see.

It is a scientific fact that what qualities we try to see in another we find. If we are looking for the noble, clean and true qualities, these come to meet our own, providing we have them. But if we show our bad qualities we will be likely to draw out theirs. If we are mean, jealous and envious, displaying our brute side, these are the traits we see in others.

Thoughts crystallize into actual things. What we feel in our heart, harbor in our mind, think about, dream about, will in time develop from these tiny seeds into full growth. Hating some one will not develop your love qualities. A revenge seed cannot bring you anything but trouble. If we want to have friends we must be friendly. If we want love, we must have love.

Whatever you send to another draws out of them the same kind of qualities. There is a law that regulates thought, which works out just as true as any other law.

In India the developed yogis say, "If a man purposely does me wrong I will return him my ungrudging love; the more evil comes from him the more good shall go from me."

The developed person of the coming day will understand how injurious are discordant thoughts, and will no more think of letting them into his mind than he would of taking some deadly poison.

Your present character is the result of the life you have lived. An adept at character reading does not have to inquire of your past history to unmask your real traits. He is able to read at a glance your thoughts and desires.

When we see a face that is all tired and sour, you can tell the owner of the face never thought of beauty and joy, but instead lived a selfish and vicious life. When we see a face that shows a pleasant and agreeable disposition, we know that person has lived an unselfish, harmonious life.

We hear people say, "Isn't he or she lucky? They have never had any trouble in their lives." It may seem that way, but the reason is that they have not let every little thing disturb their equanimity. They have lived a beautiful and sweet life. Their nature is harmonious, and therefore they are not discordant. They never speak anything but good of others, and in turn no one speaks anything but good of them. They do not make enemies, and therefore they do not have others storing up trouble for them.

Those that envy them are probably crabbed, ugly and cross in disposition. They are continually misunderstanding others and being misunderstood themselves. Their lives are discordant and they attract everything discordant.

It is impossible to harbor secret hatreds, jealousies, grudges, and to be always looking for revenge without hurting your own self. These kind of people wonder why no one cares for them. Why they are unpopular. Why their company is never wanted. They never think the cause lies in themselves. Their revengeful and ugly radiations kill their magnetism.

If you want to be magnetic and popular you must radiate kindly, helpful, sympathetic and loving thoughts, and feel friendly towards every one. Let there be no room in your makeup for bitterness, hatred or envy.

Never try to get something that does not belong to you; that you have not a right to. The coming man will realize that discordant thoughts and taking advantage of another can cause nothing but injury and no lasting benefit. He will thoroughly realize that the law adjusts everything. That it does not pay to try to cheat justice, equity, honesty, and that it does pay to be unselfish. He will do the right thing because it will bring him his just deserts—joy, peace and prosperity. We learn the meaning of the golden rule more each year. The time will come when we will think only of doing the right thing instead of the wrong.

CHAPTER XI.

VITAL MAGNETISM.

You must all realize that it would be very harmful if every one possessed strong magnetic power and used it for the purpose of gaining control over others and compelling them to carry out his wishes. Nature understands this and you have to be capable of knowing how to use power before you will have it. There are some people that have such strong mental powers that it gives them an immediate control over others. These people use their power to a limited extent, but if they knew more of personal magnetism they could accomplish a great deal more. The person with less mental power, if he studies these lessons, will be able to accomplish more than the one with strong power undeveloped. Very often I have the question asked me, "Do you think people could develop magnetism if they were incapable of using it correctly?" My reply is: "You can develop yourself a great deal, but you yourself will have to determine whether you are capable of having great knowledge.''

We all may become magnetic and be able to influence some people. Your success will naturally depend on the development of your mental organization and you will have to do all you can to conserve and build up strength and stop any habit that will weaken or impoverish you.

METHODS OF CULTIVATING MAGNETISM.

That everyone radiates magnetic fluid can no longer be doubted. That this magnetic fluid can be wonderfully increased, we know. The best methods of doing this will be found in the following instructions:

The eyes are the means by which we communicate our inner intention to another, and therefore we should see that these organs are cultivated to the highest possible extent. We use the eyes to express with and if you will make a close examination you will find that the eyes determine the first impression, whether it is good or bad. You may develop a steady, clean and a very penetrative eye by looking at yourself in a mirror. You will find there are very few persons who can look you straight in the eye while talking to you. Look at the mirror and practice keeping the eyes fixed without winking. Imagine that you actually look at another person—a person you want to affect by your influence. Not only look at your eyes in the glass but concentrate your thoughts through the eyes. I had a student one time that changed himself in a few days by means of the looking glass. He had always been in the habit of looking downward when he talked. Never looked anyone in the eyes. In my first lesson I told him to look into the mirror and smile, "Now," I said, "do you notice that you have a very attractive pair of eyes?" He said, "I see I have, but I did not know it before." That is the trouble with most people. They do not know that the eyes have numerous uses.

In taking exercises be careful that you do not strain or irritate them. Just as you feel your eyes tiring close the exercise with a quick, magnetic glance in the mirror, just as if you met some one you wanted to impress. You will find at first it will be hard to keep your eyes from winking and keeping the lids back, but you will be able to do this in a short time. In the beginning you probably will not be able to continue the exercise for over a minute, but in a short time you will find that you can look at one object for some time.

This will bring your mind into the exercise and you will find that by keeping track of the minutes there is a certain amount of interest to see how long you can keep your eyes concentrated on the second-hands without blinking. Another good exercise to put more expression in your eyes while you are talking to some one is to look steadily in their eyes and speak every word, as it were through a visual sieve. Never stare or look fierce, unless you really want to frighten them. By a little practice you can cultivate the eyes as they can be controlled as you wish them and you can focus them at one point for an indefinite time. But never over-tax or strain them.

MAGNETIC HANDS AND FINGERS.

If you can gently slap a person on the shoulder in a more artful than a bold way, and at the same time concentrate your mind upon the contact and willing that you emit a current of your magnetism to them you will find that in many cases he will experience a tingling feeling of warmth or a noticeable shock as from an electric battery.

There is no one that is not susceptible to magnetism, but there is a big difference in people. Those that have a powerful mentality could only be influenced by another of exceedingly powerful mentality.

Before you can influence another they must be in a passive or open mind. Those that you think you could least affect would be the easiest for you to affect. Never try to influence a person when they are excited, nervous or under great anxiety or mental trouble. If these conditions exist try to remove them by sympathy, cheerfulness and your assurance that everything will probably turn out all right in the end. The more passive a person is, the easier he is to affect.

The hands can be made a powerful means of imparting your magnetic fluid to others. If your hands are naturally soft and silky, and perspire freely, it is of great help, as the magnetic fluid has an easy outlet. But if your hands are not so they must be made so by rubbing them with some good oil every night. Also rub your hands together a good deal. In this way your circulation is increased.

This will soften and freshen them up, and cause the magnetic fluid to flow more freely. Whenever you have the chance of placing your hands on one you wish to influence you should do it, and at the same time will that you are imparting a flow of your magnetism to them. This will arouse the force within you and increase your magnetic power.

MAGNETIC FEET.

It will probably be new to most of you that the lower extremities are of very great importance in applying magnetic force. As you use your hands, your feet are throwing off streams of vitality which is assisting you, although you are not aware of it. There is always a powerful magnetic-current flowing from the toes and there have been some wonderful things done by people with their toes, that did not have hands. It is a matter of fact that the feet are generally well adapted for the flow of magnetism. By being protected by shoes they are naturally softer; by walking they become warmer and more moist and emit the vital fluid very easily. They do not need any particular cultivating. All you have to do is to become conscious that they are capable of imparting the force. Just remember that your feet will do their part. Light shoes with suitable stockings should be worn. Never wear heavy shoes that will not give your feet some freedom. The tighter the feet are incased, the more difficult for the vital force to get out.

CHAPTER XII.

THE LAW OF MAGNETIC THOUGHT ATTRACTION.

The kind of people that are liked.

"The universe pays every man in his own coin. If you smile, it smiles with you in return. If you frown, you will be frowned at. If you sing, you will be invited in gay company. If you think, you will be entertained by thinkers. If you love the world and earnestly seek for the good therein you will be surrounded by loving friends, and nature will pour into your lap the treasures of the earth."—Zimmerman.

It is natural for us to like sunshiny, cheerful, bright and helpful people. The grumblers, fault-finders, slanderers are never liked. We like the people that are looking for the good not the bad in anyone. People with serpent tongues, idle gossipers, temper-losers develop ugly natures and are never liked.

It is not one bit harder to be in the habit of going around looking for the good and beautiful instead of the reverse. It is just another case of it seems easier to do the wrong than the right. The great secret of contented and discontented people is that the former are looking for the cheerful and bright, and the latter for the dark and gloomy. There is always a brighter side and a darker side. Always look for the brighter side. It will make a great influence in your character, in your happiness, prosperity and success in life in general.

There is always a light behind the dark. Look for it. You will see it. Think helpful and inspiring thoughts and you will soon be looking at many things in a different light. You can transform your character in a remarkably short time.

We hear people say "I could be happy under different circumstances." I say positively that circumstances have little to do with it. It is your temperament, your disposition which makes you enjoy or not enjoy.

Think of people that are very unhappy. We can see them grumbling about circumstances, hard times, lack of wealth, when they are a great deal better off than many others. These people would think you very fortunate if you were in precisely their condition.

If you have been in the habit of complaining about your business, talking evil about others, just try the reverse. Assume an encouraging prosperous air and you will soon see a way out of your difficulties, a change in your condition.

Remember first, last and all the time, the strong positive man does not talk or think in the negative. He never lets himself feel "I can't," but "I will."

"Brooding o'er ills, the irritable soul, creates the evils feared."

THE VALUE OF CHEERFULNESS.

The successful man is cheerful and hopeful. He has a smile on his face, and meets everything that comes in the same way.

The cheerful man is creating new power, while the pessimist is destroying his power.

There is nothing that will help you meet the hard turns in the road; sweeten life and take out drudgery, like a sunny, optimistic disposition. Two may have practically the same ability, but if one is a cheerful thinker while the other is despondent, and gloomy, the former will leave him far behind.

Cheerfulness is a tonic to the mind. It dispels friction, worries, anxieties, and it will often turn disagreeable experiences into agreeable ones. You can do your best work only when you are in a cheerful state of mind. When you are out of temper your entire physical machinery is out of working order.

It should be your plan to follow the following philosophy: Try as much as you can to let nothing distress you, and to take everything that happens as for the best. Believe that this is your duty and that you err in not so doing.

The next time you are out walking make it a point to look at every one you can. See how many you see that look as though they are happy. Notice the ugly grouchy expression on most people. You can make life one continuous round of sadness, instead of joy and gladness.

How few there are who bring more sunshine into our lives, who scatter gladness and cheerfulness wherever they go. It is so seldom that we see one of these cheerful faces that every one is attracted by it. Get the habit of looking pleasant, of radiating cheer, wherever you go. This will make you happier than to own many houses or any kind of possessions. It is free to you. All you have to do is to develop it. Your ability to radiate sunshine will add greatly to your power.

It is not really hard to transform a gloomy disposition into a cheerful one. True a cheerful face is but the reflection of a big glorious heart. You cannot look a part unless you feel it.

A SIMPLE WAY OF GIVING CHEERFULNESS.

Take an interest in what is going on in the world about you. Take an interest in those you meet. Try to open up their clam-like disposition. Be friendly to all. All these will help you to develop cheerfulness. We will never find out what is good and noble in others until we find similar qualities in ourselves. Make people think they are all right, and they will think you are all right.

It is really wonderful the effect that one person with a cheerful frame of mind can have in a crowd, or gathering of gloomy, melancholy people. Let such a one enter into a room where the conversation is lagging, and where the people feel strained. Immediately all who are present change to a joyous spirit. Their tongues are let loose, and almost instantly the whole atmosphere vibrates with gladness and good cheer.

We know the need of sunshine for plants, but we do not know that sunshiny dispositions are also needed. It will pay you handsome dividends to cultivate more sunshine in your life. It is hard to estimate the immense value of a sunshiny disposition. It acts as a magnet to draw to us the good things in life.

"In every person who comes near you, look for what is good and strong. Honor that; rejoice in it; and as you can, try to imitate it, and your faults will drop off like dead leaves when the time comes."—Ruskin.

Say to yourself, I will never speak unkindly of anyone. If I cannot see something good I will see nothing and say nothing. You will notice it will make a wonderful difference in you. Rapidly you will notice how different you look at life. You will be able to see joy and peace everywhere. If you will form the habit of looking on the bright side of everything there will be little trouble to bother you.

A great many charming characters are hidden under a mask by a habit of making cynical remarks. It keeps out of sight your happy, cheerful, wholesome self.

"Overcome the fearing tide, there's a sparkling gleam of sunshine, waiting on the other side."

We do not realize that when we are talking of our business, of our poor health and complaining in general that we are attracting these very influences.

On the other hand let us think and talk the opposite, and we will attract the opposite. We will become positive, vital, magnetic.

Negative thoughts will kill ambition. It will make your life a failure. As long as you hold to these you will never be powerful or magnetic. The negative person is a slave, he slays self-confidence instead of increasing it. A man's station in life is determined by his self-faith, self-confidence. You will never advance until you think you can. Never think some one is better than you. Never place a limit on yourself. Do away with all negative thoughts. Think of being powerful.

"He can, who thinks he can," is a saying that will always be true.

The determined man makes a road over which others follow him. He laughs at barriers that stop others and he jumps over them as if they were nothing.

You want to cultivate deep conviction. Don't let everybody's opinion change you. You have a mind of your own—use it. If you let everyone's opinion affect you, you will not have a deep conviction. Your thoughts will not be backed up by your confidence. The majority of people are not capable of deep conviction. They are superficial. They are easily changed by others' thoughts. They succumb to the first argument. They are just between, never on one side of the fence or the other. They are always willing to agree with the other fellow instead of having the other fellow think as they do. There is no backbone to such people. You want to cultivate positiveness and decision in everything you do.

If you do not have the power to stand by your resolutions no one will have confidence in you. You often hear people say, "Oh I think he is a good man," but by their tone you can tell they are not wholly sure. You want to act so there will be no doubt in the mind that you are all right. You want to be able to inspire confidence in those you meet. You want them to believe that you amount to something, that you are capable of doing important things. To accomplish anything worth while you have got to have a deep conviction that you can achieve whatever you undertake. Your determination must be strong. Such a man has influence that amounts to something. He is not easily changed to be like another, who holds different opinions.

Once you are aroused to the fact that you can be what you wish to be; that you can do what you attempt if you will cultivate the power of affirmation and hold in mind persistently the thought that you are going to accomplish what you wish, you will be able to change yourself in a short time. You can realize your highest dream.

Remember this, that if your Creator has implanted in your breast a desire to do something, He has also seen to it that you have the ability to do it. On the other hand do not think of anything that you do not want to have happen. Do not let anything make you unhappy or depress you. Let nothing ruffle your harmonious nature. You have no weakness unless you think you have. It is your thinking that makes you weak. If you feel yourself getting out of sorts, discouraged, blue, disheartened, you will find you can dispel all of these thoughts almost instantly by thinking of something pleasant, a more agreeable experience, or by thinking of something that will give you pleasure. When your thought is changed your feeling is changed. No matter what your environment is, by thinking as you should you can change your outlook and make yourself happier. Thoughts of the wrong kind deplete you, but the right kind are a great tonic.

Keep this affirmation before you: All that I dream of, all that I long to be, will be within my reach if I affirm sufficiently strong, and focus my faculties with sufficient intentness on a single purpose. Whatever you wish to accomplish you must concentrate on. The jack-of-all-trades never amounts to anything. If you want to be a lawyer you have got to concentrate on being a lawyer and not upon being a doctor or something else at the same time. It is concentration that brings to you what you wish whether it is money, health, position or the love of some one. No one would ever win the love of another until they concentrate all their love on that one. Whatever you wish to accomplish affirm persistently and concentrate all your power on securing it and when your mind is positive enough it can create what you wish. We can become a magnet and draw to us what we wish.

CHAPTER XIII.

WHY WE ARE JUDGED BY OUR THOUGHTS.

"Actions speak louder than words," is an old saying, but a true one. We form our opinions of others by what they say of others, rather than what they say to us. We could not expect them to say anything detrimental to ourselves. Don't for a minute think that you are judged only by what you say, or by your past history as declared by you. One mind is capable of transferring thought to another. Therefore what you really think in your mind of yourself is transferred to others, and that is just what they think of you. The kind of thoughts you think is detected by them. Your character is judged by your radiations and you are estimated accordingly. We meet a person and we soon form an impression of him, and it is mighty hard for us to change that impression. If we should state this impression to another, they

Judged by Our Thoughts 145

might try to change us, but it would be difficult to do so. "What you are speaks so loud, I cannot hear what you say."

Whatever we think about a great deal, we become like unto; it is pictured in our features, our eyes, face, and manner. The character reader can tell the physician, the lawyer, the clerk. He can tell what has been the dominant thought of their minds. Whatever we concentrate our minds on becomes pictured in our manner, our face and our conversation. Your secrets can be read by the qualified person.

You probably never realized it is possible to read in your face and manner the record of your thoughts; that in your face, the records are written of what has been going on in your mind for years. You probably have imagined your thoughts were known only to yourself, but they are written in your makeup. Really you cannot hide anything. Our whole action speaks what we are. We color the atmosphere with our radiations. We cannot pretend to be something, we are not. People judge us by what we are, and not by what we pretend to be.

A person may appear very pleasant, agreeable and considerate of us, but if he does not feel the same way in his heart towards us; if his thoughts are antagonistic or there is an inward grudge, and he is not what he pretends to be; our instinct reveals all this to us. We see through the mask and pretense, and instead of him deceiving us he only deceives himself.

If you only knew the real opinion of others in regard to you, you would often be surprised. Many a person has tried to make a good impression and believed that they had, when as a matter of fact they had made a very poor one. In this busy world there is always some one to cast a dark shadow across you, to depress your buoyancy; to strangle your aspirations; to crush your hopes; if you let them. They will cause you misery and suffering by their cruel malicious sarcasm; ungenerous criticisms; envious and jealous thoughts; hatred; anger; desire for revenge.

The person who is down in life wants company. He does not want others to be different from himself. If we associate with good people our good is brought out. If we associate with mean, despicable people, these qualities are brought out. It therefore behooves you to be careful of your associates.

BE CAREFUL OF YOUR THOUGHTS.

Do not have fear thoughts.
Do not have doubt thoughts.
Do not have failure thoughts.
Do not have evil thoughts.
Do not have disease thoughts.
   But instead:
   Hold kindly thoughts.
   Hold charitable thoughts.
   Hold health thoughts.
   Hold loving thoughts.
   Hold success thoughts.
   Hold joy thoughts.

By this means you will bring sunshine and gladness, while the hate thoughts bring sadness and shadows.

Be one of the helpers of the world and try to lighten the burdens of others. Our thoughts bring our success.

"He who dares assert the I,
   May calmly wait
While hurry fate
   Meets his demands with sure supply."

Your Creator wants you to succeed. He placed no limitations on you. He has placed you in no environment you cannot rise out of. You are not the victim of circumstances. Somehow, or other, you are responsible for your circumstances. To think the contrary is undermining and weakening and the cause of untold failures and a great deal of the poverty and wretchedness of the human race. Your natural thought is dominion over the forces, and if you have succumbed it is on account of your weakness. Your birth-right is riches, happiness and freedom, and if these are not yours it is because instead of thinking of them you have thought of the opposite, and have attracted to yourself poverty and slavery and wretchedness. It is impossible for you to do anything when you don't think you can. There is nothing impossible when you think you can. Can you expect to succeed as long as you think, talk and act like a failure? Forget there is any such word as "can't" and "fate" and "doubt."

I can think of nothing so harmful to your success as to think you are weak; that you lack ability.

The big mass of failures are caused by people doubting their ability. Young men should be taught never to doubt anything they undertake. Doubt is an associate of misfortune.

It is useless to want to be something and all the time think you are going to be something else. What you want to be, think you are going to be. Don't entertain any other idea for a minute. You will never be any different than your conception of yourself.

Your success in life depends a great deal on the impression we make on others; the amount of confidence they have in us; but the confidence they have in us depends a great deal on the confidence we have in ourselves. Therefore we must believe in ourselves before we can expect others to. There is a certain magical air about really successful teachers, lawyers, salesmen, merchants, etc. We can learn a great deal by watching others. Every time you see a man that is a power in business watch him closely. You can learn valuable lessons.

THE CHARM THAT PLEASES.

We receive our greatest enjoyment when we please others; it is our duty to give others pleasure, to try to make the world a better and happier place to live in. When we are helping others, we are helping ourselves, as all the good we do will come back to us. You smile at people and they will smile back. You like the people that like you. You like the people that bring cheerfulness and sunshine.

The first requisite in acquiring charm that pleases, is to believe in the goodness of humanity. That it wants to do what is right, and if it errs it is a good deal the result of ignorance.

No one can make you clever or capable. You have to do this yourself. You may read all the books on cultured talk, but they will never make you a good conversationalist unless you try to improve your powers of conversation.

CHAPTER XIV.

MAGNETIC CHARACTER BUILDING.

Our great philosophers tell us the grandest thing in the world is character. But it is something that is little taught, and as a result we find very few people that have much real character. They hardly know the meaning of the word character. Parents will spend large sums of money to send their children through school and college, yet they are not taught how and why their character should be developed. A child in its young state is very easy to train if it is directed properly.

YOUR WEAKNESS.

None of us are perfect so the philosophers tell us and we know it is true. But they do not tell us we cannot be. Are you trying to perfect yourself? Do you try to strengthen your deficient faculties? You can become well-balanced, well-rounded, don't you think?

Then why not do it. We will help you all we can.

Your weakness shows more prominently than your strong qualities, because people are looking for your weak instead of your strong points. The trouble is with most people that they have weakness and don't know it. Their defects may be very small, but if it puts them at a disadvantage it is very important because it interferes with their results, and keeps them from rising in the world.

A great many people feel that they are peculiar in some respects. That certain traits of their character were inherited from their parents. Just because your parents have peculiar traits, this is no reason why you should. Of course you will have their same tendencies unless you choose to change them. If you think you are peculiar you will be peculiar. You could not be otherwise. We are just as we appear to be.

What we think in the mind we are. When you think of your defects you increase them, by worrying about them. The truth of the matter is that the majority of people's abnormalities and peculiarities are mostly imaginary, or at least greatly exaggerated by imagination. They have thought about them so much that to them they appear real. The way to correct them is to think about your good qualities and ignore your shortcomings.

If you are in the habit of thinking you are peculiar, get the habit of thinking you are normal. Affirm, "I am not peculiar. These seeming idiosyncracies are not real. I was made in the image of my maker, and a Perfect Being could not make imperfections; hence my imperfections cannot be real, as the truth of my being is real. There can be no abnormalities about me unless I produce them in thought, for the Creator never gave them to me. He never gave me a discordant note, because He is harmony."

By holding this thought in your mind persistently you will not think of your abnormal mannerisms or actions, and they will disappear, and soon you will see that you are a great deal like other people.

HOW TO CULTIVATE MAGNETIC BEAUTY.

There are people who are considered very homely yet they have so much magnetic beauty that they are great favorites.

It is an absolute fact that a girl with a very

homely face, and who has an ugly expression, can, if she is honest at heart, transform herself so she will seem beautiful to everyone who knows her, if she will form the habit of holding in her mind the thought of beauty. Not the superficial thought of physical beauty only, but the deep heart and soul thought of beauty. The kind of beauty that far surpasses any other kind of beauty is a kindly helpful heart, and a wish to brighten the lives of others; to scatter sunshine here and there; and good cheer. When this is done it brightens the face and makes it beautiful. A beautiful character cannot help affecting and making a beautiful life. We express outwardly what we express inwardly; our emotions, manners and bearing depends on our thought. If these are as they should be, you cannot help becoming sweet and attractive.

The most prized beauty is within the reach of everyone. You should be careful what is stored away in your brain. You should not let anything and everything gain entrance to the brain. Do not let anything enter that will not give you character and strength. Form the habit of reversing every unfavorable circumstance and every negative statement that you are confronted with each day. If you will do this it will render you and the world a great service.

For an example we will suppose that some one comes to you and tells you an untruth about another person; you believe it the truth; you repeat it to others; you find out later that it was an untruth; but you have probably already injured an innocent person. If you hear something about another person, think this over, but do not take for granted that the person is guilty or innocent. He may be either, but you do not know which. If he is likely to affect those whom you are not interested in, it is your duty to protect them, and to see that they do not become vicious. But you can do this without harm to the man. A person shall not be condemned before he has had an opportunity to establish his innocence.

Remember always:

"Things that do not in any wise concern you, should not receive your attention." Keep them from entering the brain. You have enough important things to store away without wasting time with unimportant ones. Control the gate of your brain and be the judge of what shall enter, and what shall stay out. See the truth and nothing but the truth.

If you have a position of trust, or if people are depending on you, you should not show your stupidity by placing confidence in the honesty of others. Most everyone is trying to undo the honest person. Poets and dreamers may write, "If you trust your fellows they will be worthy of your faith." But this you will find is not always so. Trust the average person and he will take your last cent and leave you broke. A thief will steal whether he is trusted or not. Unless you can afford the experiment, do not take any chances you do not have to.

There are over four hundred millionaires in this country that have risen from low positions to high places of trust. A careful study of their lives will reveal the reason why they forged ahead and became successful. For every one that starts in a given position only one out of twelve succeeds.

It matters not what position you are now holding, put forward your very best efforts. Those who advance with rapidity and reach a high place show that they are capable of doinp better things.

CHAPTER XV.

THE SECRET OF PERSONAL POWER

You cannot enjoy your talents until you develop and use them. The man that is active finds life interesting. He enjoys himself no matter where he goes. If he lives in the busy city, he gets pleasure out of being out in the country; at the country-side he sees the wonders and beauties of nature.

When we work we get more in harmony with the infinite. Try to get joy out of everything you do. Aim to do it the very best you can. If your employer cannot appreciate you, there are others who can. If you complain it is your own fault. You have a remedy at hand. The strong man does not complain, it is the weak one. You can alter your condition in life if you try. Instead of complaining, use this time in doing something. If you carry all your grievances with you, you have no time for pleasure. Instead of complaining of your environment, get busy and change this and improve yourself.

TO PREVENT ANGER.

When you feel yourself getting angry at something that has been said to you, or begin to feel worried about something, stop and commence to breathe deeply and you will find you can put your mind at ease. You see anger and worry are products of your mind. No one who gets angry and worries can see clearly. The mind becomes blurred.

Those who achieve things in life find life worth while. Everything you undertake to accomplish is an achievement that brings you satisfaction. Those positions that you are not successful in should spur you to increased effort. The man who has never suffered the bitterness of defeat has never tasted the real joy of victory. The following actual experience will illustrate what I mean:

Two men experienced a sudden sharp reversal of fortune. One was left with a bare means of living, and the other lost everything. The former turned tail on his troubles, and went away into the country where he determined to live on what he had saved out of the wreck. He brooded over his misfortunes; told people they had ruined his health; and though a perfectly strong man, became a hypochondriac. The other could not have run away even if he had wanted to. He summoned all his fortitude to his aid; all his optimism; all his faith; and fortune came tumbling at him on the heals of his troubles. What had seemed a disaster turned out the best thing that could possibly have happened. If he, too, had run away, he would never have had the fortune and would never have tasted the sweets of snatching victory from the jaws of defeat.

"Whatever troubles come to you, if you face them with courage they will pass by as the idle wind and leave behind the prosperity of which they are only the harbinger."

Sorrow and disaster come to us all some time in life. But there is a just law of compensation.

Brightness and happiness often develop from things that at first were gloomy and unpleasant.

THE OPINIONS OF OTHERS.

"What I must do is all that concerns me, not what the people think."—Emerson.

A great many people are over sensitive to the criticism of the world. But the people that have done the great things have performed their duties in their own way totally unconscious of other's opinions. They were not affected in the least by flattery or vituperations. They were guided by their own conscience. What this told them they thought was right. You should be better able to judge yourself than anyone else. They do not know your temperament or your mind. You know them better than anyone else. You should care nothing for another's praise. You are capable of praising yourself and deciding whether you have accomplished all you should.

The people that become depressed, who are continually thinking the world is cruel and unlovely and the ones that do not know themselves and the possible joy of their inner selves, are always the failures.

Their mind is so full of sadness that peace and love cannot enter. They have not learned that out of pity love is born; out of love, hope; and from hope, life. If it was not for sorrow we could not appreciate happiness.

SERENITY.

Most people are as changeable as the sea. Whichever way the waves go, they go. There are very few that stand like a stone wall and never are affected by the storm. You should try more to remain the same through the storm. Every little shock leaves its imprint and weakens you just so much. That rock which is continually washed against by the sea is in time worn away.

By keeping your serenity you will not become weakened by worries or troubles but you can emerge strengthened and encouraged. Man has gained this mastery by being stronger mentally.

The more serenity and the broader you make it the more you will be able to control the forces that dominate the world, and the less will troubles bother you. There are men that have lost their money and friends and yet they remain serene and undaunted.

"Retain your knowledge and your courage, and you will always be rich, and if you do not grow richer you will not grow poorer."

No matter what trials you are going through be brave; cultivate strength; and be thankful for what you have; that **your condition** is not worse. Let neither troubles nor grief shake your calm or dim your faith. Remember the grumbler and the cowards **are** the destitute people of the world.

CHAPTER XVI.

HOW TO CULTIVATE SUCCESS.

Ask ten men to tell what they think makes success and each one will have a different theory. It all depends on their mental calibre and outlook.

Success means perfection. And you cannot hope to become perfect; you cannot hope for complete success; therefore you cannot expect complete success and neither can you be a complete failure.

Your success depends upon your poised power. A man's success is usually judged by the money he has made. But it is wrong to judge success by the money standard. Rather success should be determined by what you have accomplished. To make money is all right, but to hoard up wealth is wrong. You should use money to do good in the world and not to cause poverty and suffering.

If you make a lot of money, but make no friends or do no good with your money, you are a failure. If you do not live a useful life and make a lot of money, you are a failure. It is what you think and not what others think that you judge yourself by. You are a world within yourself. You are bound by your own vision and your own knowledge. You have to render an account of yourself, when when you pass on to the Great Beyond. You will have no one to rely on.

It matters not then what other people said or thought. No one can help you then, and no one can help you now unless you help yourself. No one can make you happy but yourself. No one can succeed for you. Let no one keep you back then. You have a free will and you should not be bound by mental conventions. Do your own thinking, and don't let some one else do it for you.

Personal Power comes from mental independence. The majority of people take their thought from others. You must think for yourself and create for yourself, if you want to succeed. Be urgent if you want to reach the top.

Have a mind of your own. Think for yourself. Do what you think is the best way. All of our great men have been independent thinkers. They did not judge themselves by certain fixed standards.

You never know how much power you have until it is fully developed. Maybe you can influence only one and maybe you can influence hundreds and thousands. But you can never influence any one until you have confidence in your own power. The best way to develop your powers is to do everything as thoroughly as you can, and aim at perfection, all the time. Unless you do, you will be a failure. It matters not what people think, so long as you are confident that you are doing your very best, and that you are gaining knowledge and power. We cannot expect others with their limited knowledge to judge us correctly. Suppose, for instance, that Napoleon as a boy, in his Corsican home, had been able to unfold the scheme of his life and the dazzling conquests which he made. Suppose he had announced that he would win the throne by his own power. What would every one have said? That he was mad; and they would have tried to discourage him. If that had really happened and he had accepted their advice, he would not have accomplished what he did, and the history of the world would have been different.

Aim to associate with big-minded people It is well to listen to other people but don't let yourself be guided by them. Never cramp your intellect.

Our great thinkers tell us that the world is ours. "The limit of your powers is the limit of your own mind, which will expand just proportionally to the manner in which you use it."

As long as you believe something is right hold to that thought even if there are thousands that think just the opposite. You may be right and they wrong. Numerous examples have been given to prove this so it will not be necessary for us to do so.

Weigh criticism and listen to advice, but decide for yourself what is the best for you to do.

As long as you are trying for something better you are succeeding and just as soon as you stop trying you are weakening. Disaster you are likely to have, but meet it bravely and go on. Don't let the hesitating, who lack brains and courage, keep you back. When you shall have passed to the Great

Unknown, let no man say you were a coward. If you have courage in your heart the future will have no terror for you.

THE REWARD THAT COMES FROM POWER.

The object of all knowledge is to use it; you could have all the knowledge in the world but unless you used it, it would not help you. You must turn your knowledge to practical use.

If at present you cannot believe in your own power to make you physically great, you can acquire this in a few months by practicing the suggestions in this book. In a little while you will be ashamed of your doubts of yourself. You will discover yourself and develop your full heritage of personal power.

"I believe in myself," and if you work to carry out that belief and justify it, you will never lack the supreme gifts which life has in store for you.

"Be not afraid of greatness," says Shakespeare. "Some men were born great, some achieve greatness, and some have greatness thrust upon them." Which one will you be? You were born great. There is nothing to stop you from achieving greatness if you put forward the effort, and if you go about it in the right way you can have greatness thrust upon you. What we wish to accomplish we must be prepared to give up all pleasure and ease to attain.

You hear people say "If I only had money I would do great things." But when we look back over history, and read the records of the world's most successful men, we find that more of these were born in penury than in wealth and influence.

All of our records go to prove that the lower down a man is on the ladder of fortune, the more likely he is to rise. The reason is very easy to see. The rich man has so many distractions from his work, and he has never gone through the hard experience that develops character and brains.

Just as soon as a man makes up his mind that he wants to do something in life he changes from going with the tide to conscious activity. You can drift with the tide for some time, but eventually you will find yourself stranded on the rock. Unless you steer your boat, you will never get where you want to go.

There is always one course a little better

than another. We are capable of taking that course. No matter what the difficulties we are built to overcome them. It is strictly up to you, whether you will just drift or make progress. If you want to do the latter get your bearings and then steer your course right ahead, and stick to that course unswervingly, and power and success will be yours.

You have power within you waiting to be developed; you can permit this power to be useless, or you can utilize and transmute it into mighty energy.

What is the force that drives men to any goal? It is personal power. It is the spark of the Divine within him that gives him dominion over earth. It is higher than intelligence, as it makes him use this in the right way. It is higher than character, because it creates character. It is higher than personality as a man's personality is but the expression of his mind. It controls all these yet is above them, because the mind gives the power to see beauty and the strength to secure what we need from the hands of Fortune.

If your arms or hands or any of your parts were not used for a year, they would wither, from the want of use. A man's personal power is the same, only he never withers so that he cannot be brought back quickly. It will decrease, but by a little effort he can develop it into a mighty force by which he can make his life what he wants.

Your personal power will not do some one else any good. It is for you alone and must be developed by you. Every one has particular needs. You have the latent power for your own needs and not some one else's. If you did no have inherent genius you would never become an Edison or a Marconi; and here is where I differ from those teachers who tell the world you can do whatever you will to do. You could never be a Napoleon unless you had his brain. I teach that whatever you are by nature fitted for, you can do if you will use your power in the right way, and if you will work you will be satisfied with the ultimate results.

Let your actions be guided by the aspirations of your heart. If that is pure and noble then you have enough power within you to overcome any outside power. But remember anything worth having is worth working for. There are no short roads to to power. But there is a road that will lead you there and the reward that's waiting at the other end will repay you.

CHAPTER XVII.

QUALITIES THAT WILL MAKE YOU SUCCESSFUL.

The majority of men need something to stimulate them occasionally to do their best work. A big advertising company of Paris has the following notes on the wall:

"Do It Now."
"Do It Better."
"Do the Next Thing."
"Do It at Once and Do It Well."
"Do Your Best and You Will Never Have Cause to Complain of Failure."

Every morning at nine o'clock all the employees repeat the above mottoes, with their eyes shut.

It is well to have some such guiding principle in life, something that will spur you to do you your best. Form the habit of doing everything the very best you know how and it will not only bring you success, but will influence your character and change your whole future. Anything worth doing is worth doing well. Don't do anything in a slipshod manner.

YOUR ASSOCIATES.

Did you ever start to think what an important part your associates make in your life? Do you realize how much your success depends on your associates and friends? It is a well-known fact that your friends will draw you to their level. If you associate with people that are idle you will have that tendency. It is just as easy to form acquaintances that will help you. Ask yourself frankly about them:

"What shall I gain by knowing them?" If you cannot gain something from knowing a man it is not worth while to know him. This is a commercial age. Life is very short, and you cannot afford to waste time with people from whom you can gain nothing Aim to mix with your intellectual superiors and those that can draw forth your knowledge, and keep your mind active.

Associate with those that are worth measuring your own mind against. If you associate with people that continually waste their time, and have no intellectual force and no strength of character, you will dull your intellect, and your powers instead of increasing will diminish. There are a lot of men with noble characters who are industrious and intellectual and who are thorough sportsmen whom you will be the better for knowing. If you cannot conscientiously believe that you will be the better for knowing a man drop his acquaintance.

"Show me a man's friends and I will tell you what sort of a man he is." You will find that great men associate with men of great intellectual and personal force. They just could not associate with mediocrites.

You are your own best judge. Many people have the habit of using flattery to accomplish certain ends. No one is insensible to flattery, but you can easily decide whether it is flattery that is merited, and if it is not it means nothing. The clever man you will find is a close student of himself. He is ever watchful for his weaknesses so he can cure them. "A man is known only to himself and God."

If we receive praise for something we have

done that is not our best work, we should feel ashamed rather than complimented.

Don't wait for some one else to find fault with you. Look for your faults and remedy them. Recently a very successful man was negotiating with a firm and he said to the manager, "If I do not satisfy you, you will never need to ask me to go. I shall go myself before that is necessary." A man like that does not wait for another to either praise or find fault with him. He is watching his actions. He knows better than anyone else if they are up to the standard. If he is doing the very best he is capable of, he is satisfied. But if he is not, he blames himself. Be your own judge, and don't rely on what others tell you or think of you. You know what you are capable of doing and let this be your very best, your most perfect work.

The habit of doing your best work will gradually grow on you like any other habit. You will not be satisfied with anything but your best work. It will develop powers of will and spur you on to do better and better work. In the present age the man who succeeds is the one that is ever trying to do better than his competitors. This applies to the head of the firm and down to the office boy. The head of the firm wants his work the very best so he can get some of his competitor's business. The office boy wants to do his work the very best he can, so he can secure the next higher position.

A big employer was asked at a recent convention what he thought of the chances of a young man today, compared with those of twenty-five years ago. His reply was: "I have often heard it said that there are not as many chances for young men to rise now as formerly. I do not agree with this view. I believe that there are even greater chances for a young man than ever before. But these greater opportunities demand greater qualities—qualities that can only be acquired by an increased devotion to study—to greater self-discipline, and to an unconquerable determination to master the principles that underlie the profession or the business he is engaged in. Less opportunity for getting on! Why, one of the greatest difficulties of the large employers is that of finding thoroughly capable men to manage the various departments of their enterprises. There are many who think themselves capable, but few who can stand the test."

It has been said that "knowledge is power," but a man may have a great deal of knowledge with very little wisdom. Wisdom which is distilled knowledge, is undoubtedly a powerful factor in human affairs. And happy is the man who possesses it. Knowledge is no longer a step which few climb. The opportunities for acquiring it are so many and various that to be ignorant is quite unpardonable. It has been truly said that "experience keeps a dear school," but it is the only one fools will attend. Happy is the man who is always prepared to avail himself of the experience of others.

I know of a man who never made even $700 a year up to his thirty-fifth year. In three years' time it increased to $10,000 a year, and at the present time he is making over $50,000 a year. He was asked to what he attributed his success, and replied: "You know for fifteen years I was only an obscure clerk. One day I said to myself: It's no use trying to do just the same work as other people. I must do more." He began to study system, to plan his work ahead. Soon he was able to do his own work in a short time, but he was not contented to loaf during the remaining time. He commenced helping others do their work. He developed such a capacity for work that soon they were doing a great deal more work than ever before. He discovered all kinds of possibilities that would not only increase the amount of their work, but also the quality. He was soon advanced from clerk to general manager of the branch office and then was taken to the main office.

There is no question that a young man has as good opportunities today, but he must develop his powers and put them into his work. Carlyle says: "Our grand business undoubtedly is not to see what lies dimly at a distance, but to do what is clearly at hand."

If you will take an interest in your work, and do it faithfully and as much as you can, your opportunity will come. It is the lazy, shiftless kind that complain of their ill-luck. What you call ill-luck may be turned into valuable experience if you let it teach you. Your experience should bring you knowledge and this will surely increase and give you wisdom, and wisdom can bring you everything you desire.

CHAPTER XVIII.

HOW TO PROTECT YOURSELF AGAINST INJURIOUS THOUGHT ATTRACTION.

Never expect something you do not wish to happen. Expectancy is a powerful magnet. By your desire you attract what you want. If you fear something and think about it, you attract it to you. The law works for and against you unless you know how to control the forces. If you are thinking about something that you don't want to happen, don't you see you are attracting this just the same as if you desired it?

Many teachers tell you you can do away with fear by repeating, "I'm not afraid," not realizing that you are thus admitting that you are afraid. A better plan is to say, "I am full of courage. I am able to protect myself. Why should I be not as strong as any one else? Why should I think that some one can hurt me by his thoughts or action? I am positive, not negative." If at any time you feel inclined to fear, say over and over again "I have courage—courage." You cannot say this and think it and yet suffer from fear at the same time.

THE POWER OF SYMPATHY.

By means of the wireless we are able to transmit messages across the ocean. The messages are recorded in an instrument that is in tune with it; otherwise the message could not be received correctly. There is no question that the human mind is a more delicate piece of mechanism than man could ever create. But he cannot record the message sent by others. True, there have been people that were in tune with each other who could receive each other's thoughts to a limited extent.

When we have this sense developed, which it surely will be, it will be of more practical use than some of our present senses.

It is not so many years ago that we did not believe we could transmit messages through space without wires. The same forces were ready to be used for hundreds of years before, but no one had gained knowledge of how to use these forces. If two machines can be so perfectly adjusted that they can transmit messages across thousands of miles there is no reason why two human brains that are so delicately constructed could not do the same. We have evidence to back up this theory. I could cite a number of cases that I have investigated where people were so perfectly in sympathy with each other that they know if the other is sick, no matter how great a distance separates them. You have most of you read or know of such cases, so I will not take up space with them. Have you not been speaking to some one and the same thought is spoken simultaneously? Those that know each other very well can often detect what the other is thinking about. All this only proves that you have the power to transmit thought, if you only know how to develop it. I expect to spend a good deal of my time in the future in studying along this line, and some day the results of my experiments will no doubt be published. You certainly have the powers and they certainly will be developed into usefulness in time.

In order that living instruments may be used to transmit messages by wireless, they must be perfectly in sympathy. Then this must be the case if two persons wish to send messages to each other. We know that there must be sympathy before we know what is going on in another's brain, and to make our own, perhaps, felt. You cannot sympathize with another's sorrows until you know what his emotions are.

The value of sympathy consists in harmonizing two persons so each one is susceptible to the influence of the other's thoughts. If you want to draw people to you, you must cultivate sympathy. A sympathetic listener will have people open their hearts to him. There is not a question that if our brains were attuned exactly to each other that they could transmit messages the same as the wireless instrument. When two minds are in perfect tune there is a double force at work. You have both the power of the spoken word and the projected thought, which will surely reach the other's brain. You hear people say, "We thoroughly understand each other." And they do. Their minds are so perfectly attuned to each other that they understand each other's meaning before the words are expressed. They know what each other will say.

You can easily see that sympathy, if it is genuine, must exercise a very beneficial influence among friends, and even in the practical affairs of life. By sympathy you are able to secure a man's complete thoughts, and this must be to your advantage.

The person that is shy or very reserved may hold back part of the thoughts that he would like to convey, but if you can fathom what he did not say it will certainly help you, and you certainly can do this if you have sympathy enough of the right kind.

"Sympathy is the key that unlocks the door of every heart."

Those that you can sympathize with are good for you to know. If you have no sympathy you cannot give back what another gave you, and therefore that person will not be attracted towards you. If, on the other hand, you meet some one whose sympathy seems blunted, in whom you feel no corresponding sympathy, you can feel assured that the person will not be one worth while knowing, as you have nothing in common.

Without sympathy your friendship will

never be of the inner kind. You will never understand the potent forces of the human mind without this. It is capable of solving a riddle that otherwise you would never understand. Sympathy, like everything else, can be cultivated. It does not belong to just a few. Every one has some, but it is more highly developed in some than in others. It is worthy of cultivating and will help you in becoming magnetic and attracting friends. The physical aids in developing sympathy are in the voice and the eyes. The sympathetic force is transmitted by a soft-toned voice and a clear gaze that speaks understandingly. When you are talking to a person try to feel like the person you are talking to feels. We do not mean that you should accept his thoughts unless they agree with the way you think, but put yourself in his place and see if his facts or views are not just.

Do not reject his views until you have thoroughly understood them. You will find that this will greatly broaden your mind and help you to understand all kinds of people. "The Golden Rule of life is to make allowance for every one but yourself." If you do this you will find that others will make allowances for you.

No one was ever happy without sympathy. If you give it to others, you will receive it in return.

"Laugh, and the world laughs with you,
Weep, and you weep alone."

One of my friends is, I think, the happiest man in the world. His motto is: "You can get something good out of every person you meet if your sympathies are large enough." Think this over and see if it will not help you. It has helped me.

It is our duty to try to be bright and interesting and cheerful. In pleasing others we add to their pleasure and to our own. Of course, in order to please others you will have to study them. When you learn how to please others you will know how to manage them, and, learning how to manage others, you will learn how to better manage yourself.

There is an occult law that pays us back in our own coin. If you are a distributor of hope, trust, sympathy, and brightness, you will receive these in turn. Your life is not an independent one, but one of a conglomerate whole. It was never the Divine plan that you should stand alone. You need help and encouragement from others. They need your help and encouragement.

You can strengthen your self-reliance and overcome your weakness by associating with others. Whatever you are in need of, if you will give freely to others you will receive them in return. You can learn something from every one you meet. It is your duty to learn all you can, so that you may give the knowledge to others. One individual depends more or less on another.

Kipling expresses very well in the following your correct aim in life:

"Help me to need no help from men,
That I may help such men as need it."

Only the wise can help. We can become wise by sympathizing with each other and drawing out their experiences by studying their experience, and thus learn what will be practical for us.

CHAPTER XIX.

HOW TO MAKE YOURSELF A GREAT POWER IN THE WORLD.

You have in you that same divine, creative spirit that was in all our greatest men. The only difference between you and them is the difference in your mind power. You can develop this and be raised above your fellows, above your present superiors, above the disadvantages of your environment. The great spur that makes a man want to do things comes from within.

You must train your mental forces all the time, until you awaken that wonderful power within you and put it to practical use. Mind is the creator of all that is made by man's hands. The greatness of your brain determines whether your work will be superior or inferior.

Say to yourself, "I hold within my mind, in an equal or lesser degree, similar powers to those that Shakespeare had. I can raise myself above my fellows, out of the ruck of mediocrity, above worries and disappointments, by developing my mind. I can succeed in anything I plan, and I can plan noble deeds if my mind is big enough and broad enough. A little thought and care every day, a little trouble, a habit of thinking systematically and logically will, bit by bit, strengthen my mental faculties and reveal undreamed-of possibilities to me. 'I am the master of my fate.' I will develop this divine gift of mine. I will cultivate this divine soul, and I will make the deserts of my life blossom like the rose."

The greatest work is done by conscientious attention to the smallest details and requires a great amount of labor in seemingly trivial matters.

Success sometimes comes so slow that you may think it not worth the effort. When you get to thinking this way, remember all men have thought the same.

Whatever you wish to become keep it before you. Make your motto:

"My mental training will enlarge my brain powers and enable me to do great things."

Whenever you are pursuing knowledge you

are not wasting your time. The more you know the better you are qualified to cope with the unexpected. The more information you have, the more rseponsibility you have of creating new ideas.

The more you train your mind the more you will be able to make use of the information you have stored away. You should form the habit of looking ahead; of planning what you want to accomplish. You may think this all nonsense, but it works out, and all you want is results.

The calling forth of this determination and perseverance will greatly strengthen your character and will do a good deal to make you a greater man.

You should read the lives of Napoleon and Cromwell and see how hard they worked when they were so near failure, how they achieved victory through the force of their mental powers. Reading the lives of victorious men will give you the encouragement to persevere in developing yourself. Such books inspire you with noble thoughts and lofty ideas. It wakens in you desire to do things and helps you to raise yourself above the mediocre people around you.

DON'T BE A DOUBTER.

To the doubter everything seems impossible, because—

"Our doubts are traitors,
  And make us lose the good we oft might win
  By fearing to attempt;
  But on the other hand,
  Nothing is impossible to a willing mind."

If you want to succeed in anything you must have confidence in your own powers. That is the secret of those who rise. Your mind is made up of the material you gave it to digest. If you feed it on doubt, you are bound to have a doubting mind.

From doubting something you gradually get to doubting everything you do. This kills success and self-reliance. From today on forget the two words, "doubt" and "fear." There is nothing you will ever be called on to do that you have not the power within you to do. The medical professors say that the human body is never given more pain than it can stand.

If the suffering is beyond human endurance the person loses his consciousness. This is the same with the mind. It is not asked to do something that it is not capable of accomplishing.

We told you above if you give the mind doubt to digest, it will be a doubting mind. It works the other way. If you give it hope, it will be a hopeful mind. Doubt has no more place in your mind than poison in your mouth. Cast it out instantly if it tries to enter. Never say or think "I can't do this" or "Can I do this?" but always "I can do this—I will do it." You will find this kind of spirit will make you succeed.

The person with a strong mind is able to concentrate on his work with such enthusiasm and power that he will conquer all difficulty. Positiveness creates confidence. You take up difficult problems and succeed, where a doubting person would not attempt them. Success makes the mind grow. The more you undertake and accomplish the more your mind expands. If you do not believe you can accomplish your object, you do not have that pointed attack and your effort is aimless. You would not become one of the leaders, but one that is led by one of the leaders with a strong mind. As long as you have to be led you have not a free will. Assert your free will and you can become a leader and director of your destiny.

THERE IS ALWAYS A WAY OUT OF DIFFICULTIES.

There are very few people who do not at times feel inclined to lose faith in themselves, especially if things continue to go wrong. Most men in business experience, at some time in their life, bad business, when prospects look hopeless.

Then is the time that shows if you are a real man. The man that fights on, succeeds. It is at just such a time that you need friends who can show you how to emerge from your difficulty. If you have no friends equal to the task, you have enough reserve power to do it yourself. A high bishop once said, "No man is a failure until he gives up." A brave man is equal to overcome any position. Remember the old saying, "God helps those who help themselves." Those that know are sure that this is correct. There have been times that I have found myself in such a position that there seemed no way out, but I knew there was a way and, sure enough, **a way** opened up, and by experience I have **learned** that the following is true: "That, however hopeless a task may seem, if you but carry it on, using your faculties to guide you to safe methods, the way out will appear to you."

This is an expression meaning that you must go on in your business career, through all doubts, through all difficulties, and through all despair.

KNOWLEDGE GIVES YOU POWER.

There are some things you must do if you want to make your life the highest success, and that is, you must be systematic, industrious, and have knowledge. You must use all these to achieve your best efforts.

Success is governed by certain laws. No business will be a great success until some one with knowledge creates it. Then industry enlarges and retains, and then there must be system so the organization will run smoothly.

These same rules apply to man. He cannot succeed until he uses all these in his daily work.

Many may think that to attempt to define what makes success is foolish, but the successful business man knows from experience that success is possible to every one if they develop knowledge, if they are industrious and systematic. The wise man is continually studying other competitors' weaknesses and their strong features as well as his own. He succeeds because he avoids the mistakes of others and takes advantage of their knowledge.

CONVERSATION AND ORATORY.

The words you speak are the most powerful weapons in dealing with others. By your words and your eyes you impress, either individually or your audience. You use your eyes to convey impresions to the brains of others, and they use their eyes to convey impressions to your brain. You need to cultivate both of these things.

The majority of people practically do not use their eyes in conversation, but if you wish to make the best use of your mental powers it is necessary for you to do so. But you don't want to gaze at another so fixedly that he feels uncomfrotable. You want to create a pleasing effect, and not lose it. Get into the habit of talking straight at the person you are addressing, so that he feels that you have his whole interest and attention. That is what a person wants, and that is what you want to convey. The person that is talking to you does not want you to be looking around the room. He does not want you to attempt to read something at the same time. He wants to look at your eyes, as they convey to him what is going on in your mind.

YOUR FOUR EYES.

Did you know you have two pairs of eyes? One pair is in your head and the other in your mind. The vast majority use only the pair in their head. They are really blind to their other pair, and as the result they see only one half of the thing they should see.

You express your emotions through your eyes. Whenever your smile is genuine your lips laugh with the eyes. By studying these two you can tell which it is. When you are sorry your eyes show it, and the person that is a student of human nature can tell whether your eyes reveal truth or falsehood. You can show the interest in whatever you are doing by your eyes.

When you meet a person be careful to look at him or her directly between the eyebrows, and keep this up while shaking hands. You must lean forward two or three inches toward the one you are shaking hands with or speaking to. The right foot should be about twelve inches in advance of the left one. A drooping of the eyelids or a downward look shows that the person has been influenced by your personal magnetism. Never let him or her look you out of countenance.

The theory is that when you look at a man between the eyes it appears to him that you are looking directly into his eyes.

You cannot exert the charm of your intellect on another person unless you have a good command of speech. Every one should cultivate the gift of conversation.

You obtain your effects in conversation in two ways. You tell another what you are by your voice and your manner, which either charms or repels. The way you speak depends upon whether the other person will believe in your conviction of truth and what you say. The culture of your mind is revealed in your conversation and what you speak about. Whatever you talk about you should know thoroughly; it is no use for you to have a great fund of knowledge unless you can convey it to others in a pleasing way. If your voice is harsh it will take all the charm away from the most beautiful sentence.

Affectation should never be indulged in. It should be your constant aim to be natural and cultured in your conversation. Some people think by shouting they are more forceful, but you can be just as forceful when you speak in a quiet tone, and you will show off to better advantage and be more pleasant to listen to. You can develop your real conversational ability and learn from your conversation a great deal, as it is a great educator. One of our professors gives the following to develop yourself:

Never say anything about anybody which you would not wish them to hear. Never say anything that will hurt the susceptibilities of the person you are talking to. Never be afraid to utter your opinions and to stand by them. Never miss the chance of saying a kind word either about absent friends or about your listener.

If you will bear these points in mind and try always to be interesting, even at the family table, you will soon become a proficient conversationalist. When you have acquired the art of talking well, practice diligently the more difficult task of listening well, and you will become a perfect talker.

CHAPTER XX.

PERSONAL MAGNETISM PREVENTS DISEASE.

Promotes health, ambition and cheerfulness, which is real life.

I want to firmly impress on your mind the advantage of starting the day right. On awakening in the morning, without starting to think about it, unconsciously feeling that you are doing right, that your course is a successful one, stand before the open window where the fresh air is coming in throw back the head, raise the chest, throw out the arms, draw them upward and above the head until they are rigid, and take in six deep breaths of pure oxygen. Then relax; rest for a few seconds. Repeat this three times, and then you will feel full of vital force, and you should smile. You will notice the eyes have more of a sparkle and radiate brilliancy. This is caused by the blood filled with fresh oxygen coursing tumultuously through the veins.

You will feel like the birds in springtime and that life is glorious and your wildest dreams are possibilities.

To accumulate personal magnetism proper attention must be given to breathing and tensing the muscles through the nerves. When you make the muscles rigid they are charged with magnetism which has been sent over the nerves from the brain. Practice long, smooth and deep breathing, and accompany this with the tensing and relaxing exercises given in my previous work. You will find in a short time you will be able to feel the magnetic thrill through every muscle and nerve of your body. After you have exercised for a few minutes, and after you have completed the exercises, you will feel more vigorous. You will never find a magnetic person sickly, as magnetism is a great prevention to disease. It will keep you from having colds and catching contagious diseases. The person who is tired and sickly all the time could change himself in a short time by the cultivation of personal magnetism. The magnetic condition is a warm, active and generative condition. The new force that you will instill into your system drives away the old worn-out force. It is the motor from which you receive more energy and more vitality. Now don't waste it, but use it to the best advantage.

Most people fail because they don't devote enough consistent work to the head and the brain. Get control of this and use it in the right way. Study every voluntary act of your life and see that these are done in accordance with a definite judgment or purpose, and be determined to accomplish this. There are two qualities that are very essential in learning the power of influence, and they are: Purposeful judgment and determination. The former is the result of concentration and the power to concentrate your mind on a definite object with an assuring belief in the fulfillment of your desire. This is the secret of Person Magnetism. You must breathe into those lungs of yours, so you will imbibe in the blood the life-giving properties of the air, and do this with the mind concentrated on the belief that you are going to receive, and you will be given, new life, hope, ambition, which all go to make up personal magnetism.

The secret of success in anything is the repetition of the ordinary things of life.

Don't neglect your exercises for a single day, but be constant in them, and you will have bodily perfection. After once you have formed the habit of taking them regularly you will feel more keenly the vital or life force running through your body, and as you continue them you will not only get the benefit of the present one, but you get an accumulated benefit.

SPECIAL EXERCISE FOR DEVELOPING A MAGNETIC PERSONALITY.

There are come important exercises that will help every one to become magnetic. These will make you beautiful in mind, in body, in character; they will assist you to attract the opposite sex; they will help you to secure love and retain it; to attract friendship and the world's esteem.

The correct position, whether you are walking or sitting, must receive your attention. You must have your head up, the chin drawn in and the chest expanded. Always have the shoulders back and down. The chances are you will find this a little hard at first, but soon you will find it just as easy to haintain this position as it is your present one, that the chances are is a bent-over one. All it requires is persistent effort for some time until the unconscious mind takes it up and makes it your regular position and a part of yourself. Be particular about this, as it is very important. See that your position is correct and forget you ever had any other kind. Whenever you are out in the open throw every bit of air out of your lungs that you can and breathe by long, smooth inhalations. Draw in the abdomen and swell out the front and sides of the chest, and then exhale this with an even smoothness and steadiness. Keep up this way of breathing until you have acquired it and it becomes as natural as breathing at all. Don't do this at intervals, but do it all the time. Make this your way of breathing and do not forget it.

You have now been taught some mighty good habits of standing, sitting, walking and breathing properly. These habits should be by now your natural ones and a part of your real life. You now are ready to make the whole body magnetic. I told you previously that when a muscle becomes rigid you are making this magnetic, and now you want to make your whole body magnetic. You can do this by faithfully following the following rule implicitly:

Stand, holding yourself in your new position, and inflate the lungs at the same time; slowly raise the arms straight out and up from the sides; continue inhaling nad raising the arms until, when the lungs are completely filled with air, the hands touch above the head. As you slowly empty the lungs, bring the hands down to their first position. Now keep the arms tense, as this will magnetize the muscles. This exercise should be continued for at least five minutes four or five times a day.

Stand in the same position and extend the arms out straight in front of the body, and slowly bring them back past the sides until the hands almost meet behind the back. Now, as you carry the arms backward fill the lungs with a long, smooth, deep breath, and as you return the arms, exhale. Repeat this exercise ten times slowly. Raise the arms forward and upward from the sides, while exhaling, until the hands meet in front of the forehead. Fill your lungs and hold your breath while you swing your arms backward, downward and upward, and then up again to their position in front of the face, after which you slowly let them come down to the sides while you exhale the air. Do this several times.

You may think these exercises have nothing to do with magnetism, but if you will practice them for two months you will see what a great deal you are benefited.

Now for some more important exercises. Lie flat on the floor with face downward, arms by the sides. Draw the arms slowly along the floor in a circular sweep until they are extended in a horizontal position at right angles with the body. Stand with heels together, but your weight on your toes. Now throw the head back until it rests on the neck. Exhale all the air out of your lungs. Now inhale gradually, by the method you have learned, and raise the body by drawing the arms slowly inward. Let the palms of the hands be flat upon the floor and gradually draw them in, keeping them at right angles, in a straight line, towards the place where the body touches the floor. This movement raises the body from the floor, and when the arms have reached a point exactly in a straight line forward from the chest, the body is raised, or rather the upper part of the body, the length of the arms above the floor, while the lower part of the body is upheld upon the toes. Now, keeping the hands at the same point upon the floor, bend the elbows, letting the body down until the chest (the head is thrown back) touches the floor. Raise the body again, inflating the lungs, and lower it while exhaling the air. The chances are you will not be able to do this over three or four times in the beginning, but gradually you will be able to increase the movements to ten.

If you will continue these exercises for six think of stopping them. The truth is you will months, you will be so benefited you will not be so different in six months that you will hardly recognize yourself as the person who started the study of these lessons. You will stand correctly. You will look back over the time you have spent practicing the exercises with a smile and you will say it was time well spent, because you will then realize that you are rapidly acquiring that great, wondrous attribute the possession of which is the secret of all greatness—Personal Magnetism. These exercises will not only develop the muscular fibres of the body, but you will have also accomplished something greater. This is developing the nervous system. You will never find a nervous person that is magnetic. If you cannot control yourself you cannot influence another person. It is utterly impossible to develop a magnetic personality if your nervous organism is not under complete control.

The majority of people waste a good deal of energy by unneccessary movements. Some have the habit of tapping the feet on the floor, drumming on something with the fingers, humming, whistling, etc. All of these show lack of control, and until you stop these mistakes you will not have control enough to affect others. Watch yourself carefully and see how many useless movements you make. Keep on your mind the necessity of having perfect control over the features, hands, feet, and every organ of the body. Be careful how you sit down; breathe properly and let the mind become imbued with the idea of calmness, coolness and passivity. Dwell on the thought that you must control your body completely in every motion. By thinking consciously for a time the subconscious mind will take it up and overcome it. It is an excellent idea to sit a couple of times a day in the correct position, keeping perfectly still, without thinking of anything. Let not a single thought enter your mind. See how long you can sit still without a twitching of the features, swaying the head or any portion of the body. In fact, avoid all movements of the voluntary muscles. Practice this until you have perfect control, and this is a great force in personal magnetism. When something happens that has a tendency to worry you, just think "This is just a little speck on the horizon of my mind and I will brush it away." Acquire a passivity, a self-confidence, a self-reliance, and know and believe that you have but to dismiss the worry to obliterate the origin.

Make up your mind firmly that you will not let anything ruffle the placid surface of your mind.

When you have mastered all your emotions and can sit, walk, stand and talk for a long time without any unnecessary movement of the muscles, you have acquired perfect nerve control and are a great deal nearer to the top of the road that leads to self-supremacy and success.

It has well been said that the eye is the most

potent and powerful factor in the ability to influence others. There is no other organ of sense that emanates such an expression of magnetism and intelligence as the eye does.

There are a few simple rules for developing the nerve center of the eye, which I now give you.

Fix a mirror so you can look straight in it. Put a small dot on the glass about level with your eye; fix your eyes on that dot, keeping them there while you slowly move the head to the right as far as you can and still see the dot, and then to the left as far as you can. Repeat this several times every morning and night for several weeks. These exercises strengthen the muscles of the eye, and also of the eyelids. After you have done this for several weeks slowly you can increase the speed of the motions. It is important that you turn the head as far as you can to the left and right, but you want to be able to see the dot all the time. You can now lengthen the practice and continue the movements until you feel a strain or an aching. Always stop just as soon as there is the least bit of strain felt.

Now stand in the middle of your room and

focus the sight on a point in the center of the opposite wall and then slowly raise the eye to the top of the wall, then down to the floor; now to the right side and then to the left, always crossing the center point when going from one point to another. Make the motions slowly, systematically and faithfully. Continue this about two minutes each time. You can gradually increase the length of time to exercise. Your eyes should be strengthened enough to take up the next step. Stand before your mirror in exactly the same position as the first one and focus your gaze on the pupils of your own eyes reflected from the mirror with an intent, unblinking gaze. Now open the eyes to their widest extent, keeping the pupils focused, and gradually close them—not altogether, of course—but as far as possible to permit of a retention of the gaze. This should be kept up until the eyes begin to fill up with water. You can lengthen the exercise after a few days.

After practicing this exercise until you can do it five minutes at a time, you are then ready to take up the next one. Stand before the mirror as before and gaze into the depths of your own eyes, and look as though you were trying to see some object behind or beyond your eyes. You should stand about ten feet from the mirror while doing this exercise. After practicing for two days take one step nearer the mirror each day until you are close to the mirror. Keep the pupils always focused on each other. Don't think you are looking into your own eyes, but that you are actually gazing in some one's eyes. This will require a good deal of practice, but it must be kept up to cultivate a magnetic eye. You must practice all the exercises daily. It is the regular, persistent work that succeeds.

After you have practiced the exercises daily for a month you should be able to look any one in the eyes with a compelling and magnetic gaze. You are not to stare at any one in any sense. Your aim is to acquire a penetrating gaze which gives you a complete mastery over those with whom you come in contact and makes them unconsciously acknowledge your superiority.

THE POWER OF ATTRACTION.

The sooner you get the idea out of your head that power is inherited the better it will be for you. You alone work out your own destiny for good or evil. No one else can give you magnetic force. This lies within yourself and has to be developed by you. The so-called geniuses of the past and present, the great powers of the world, the strong personalities of all times have been those that have cultivated their magnetic force. It is true that they inherited tendencies of their characteristics and qualifications, but if these had not been developed they could never have given the world that which the world has never known, so in a sense they did not inherit them.

It may be that their wonderful powers were developed with less conscious effort on their part than is required from most people, but you all have wonderful powers if you will only develop them. The trouble with most people is that they have too little desire, or desire that is so weak that it does not arouse conscious effort. Your will needs development. You want to realize that you have no limitations and that you are not an automaton that is run by forces over which you have no control. If you believe in this way you will never have the motive force and impulse that would bring a realization of your ideals and desires.

The secret of doing things is to believe that you can do them, and you will. The object of all my lessons is to teach you that you are a positive, conscious and intelligent center, a magnet whose drawing power is expressed by his desires, and his desires are simply a realization of his ideals. If you want to be a power, you must first have ideals, then set about to realize them and keep striving until you do.

It is very important that you realize that you want to possess a powerful magnetic personality. You must, once and for all time, dismiss all idea of limitations. The height you will reach depends on yourself. When you have cleared your road from limitations and have substituted perfect reliance in the realization of your ideals and desires, and thoroughly recognize that all power is within you, you will have entirely and forever banished poverty, fear and disease, and will develop your personal magnetism to a degree you never dreamed of before. If you want to make something of yourself out of the ordinary, you can by studying these lessons. Ask yourself this question: Shall I do this? There should be but one answer, "Yes."

CHAPTER XXI.

A FORMULA FOR CREATING HAPPINESS.

Here is the secret of happiness and an antidote for worry poison. This is a formula that will work just as well for the young or old, rich or poor, weak or strong. It is a formula that took me many years to compound and I want you to consider it very precious and never forget it. You will find it will always help you.

It is a formula that will cure any one of worry. It is of wonderful value to all. It is:

DON'T FORGET TO LAUGH.

When you laugh the heart beats faster and sends the blood bounding through the body. There is not a tiny blood vessel throughout the whole body that does not feel the waves of motion of a good hearty laugh. Laughter increases the respiration and gives a glow to the whole system; it will brighten the eye and expand the chest; it will force the bad air from the tiny cells and will do a great deal to help you keep your health or restore it.

Grief, bad news, worry, anxiety, fear, destroys your poise, while laughter restores it. Remember this. Laughter will help your digestion. That is why you enjoy eating so much better with pleasant company. Public speakers understand this, and that is why, when called on to speak at a banquet, they try to tell something funny. Laughter supplies the brain with cheerful thoughts and, as you cannot do two things at the same time, you cannot worry while you laugh.

If you are not in the habit of laughing, get the habit. Read funny stories once in a while. Try to tell them to others in a funny way. They will make others laugh, and you will find that by making others laugh you will get the spirit and laugh yourself.

Laughing is really a tonic, and for many things it is better than a doctor. The following true story will give you an idea of how valuable it is:

"A woman had a crushing sorrow; despondency, indigestion, insomnia, and other kindred ills followed. She determined to throw off the gloom which was making life so heavy a burden to her. She established the rule that she would laugh at least three times a day whether the occasion was presented or not. She trained herself to laugh heartily at the least provocation, and would go to her room and make merry all by herself.

"This woman was soon in excellent health and grand spirits, and her home became a sunny, cheerful abode. At first her husband and children were amused at her, and, while they respected her determination because of the grief she bore, they did not enter into the spirit of the plan. But after a while the funny part of the idea struck the woman's husband, and he began to laugh every time his wife spoke about it.

"When the husband came home he would ask her if she had taken her regular laugh during the day, and he would laugh when he asked the question, and he laughed again when she answered. The children thought that mother's notion was very queer, but they laughed just the same. Gradually their children told other children, and they told their parents. The husbands spoke of it to their friends, and finally the neighborhood people when meeting this woman would ask her how many laughs she had today. Naturally they all laughed when they asked the question, and that made the woman laugh, too.

"This woman had been suffering from the greatest kind of sorrow, but the simple act of laughing three times a day brought her out of it and put her into a new way of living. It relieved her of indigestion, banished the headaches, gave her poise and peace, and her whole home was a much better place for all the family. The entire neighborhood received benefit from the good suggestion." It became the most happy little town imaginable. Others that suffered from the same complaints that this woman did, took up the laughing habit and were benefited. The druggist's business decreased in drugs over fifty per cent in three months.

If you will follow the suggestion of laughing three times a day you will find out how beneficial it will be.

The next time you feel yourself becoming angry, force yourself to laugh, and you will find that instantly your anger disappears. The other fellow cannot get real angry with you unless you get angry. The next time some one gets angry with you, retain your poise and smile and see how much better you are able to adjust your difference.

The Bible says:

"He that is slow to anger is better than the mighty; and he that ruleth his spirit than he that taketh a city."

"A merry heart doeth good like a medicine, but a broken spirit dryeth the bones."

Always try to keep in good humor, for anger kills happiness. Do not quarrel; remember that "a soft answer turneth away wrath." Let the other fellow lose his temper and get angry, but you keep calm and you will finish on top.

CHAPTER XXII.

THE MAN AND WOMAN THOU WERT INTENDED TO BE.

"Many omniscient men and women will soon walk the earth. Omniscience and Freedom are the goal of all, and in this Great Age of Light many Egos are approaching the blessed omniscience state."

Man is gradually developing a well balanced, symmetrical mind, and behind this he is thinking more of his Great Creator and has more faith in His powers.

You are part of that great mind that creates and governs all things. There is nothing that can rob you of your birthright. The closer you come in contact with that great power the more power you will have.

Only the ignorant think they are the products of chance. All nature works in accordance with certain laws. There is never any reason for anxiety, fear or uncertainty as to what will happen. By continually expecting something to happen you bring on yourself that feeling of uneasiness, which spoils your poise and upsets your equilibrium, which is very fatal to your success. You are part of one great mind and are inseparable from it.

The coming man of the higher civilization is going to be so ambushed with thoughts of health, joy, harmony, peace, gladness, that he will have no room for discordant moods, accidents, misfortunes, etc. The coming man will be able to make himself one great magnet for attracting only those things which will increase his prosperity and happiness. He will not be sick, because he will live in harmony with nature's laws, and he will be able to change a disease thought to a healthy one.

He will be cheerful, as he will keep in stock only thoughts that will produce happiness. His mind will be closed to anxiety, worry, jealousy and envy. He will let the others that don't know mourn; he is going to rejoice.

He will never think of allowing thoughts of pessimism, wretchedness or discord to enter his mind, any more than he would think of taking poison. He will be just as careful in choosing his thoughts as he would be in selecting company to entertain if he had a minister visiting.

The coming man will know there is no such word as "can't." The man that says "can't" is a doubter.

In the new man there will be no doubts. He will have no fear, which is most people's greatest enemy.

He will have developed himself to such proficiency that he will know he will be capable of meeting any exigency which may arise.

The man of the future will never entertain a poverty thought, but instead will think prosperity, dream prosperity, and will not limit himself. He will attract abundance.

The man of the future will only live in an atmosphere of love and joyousness.

He will be healthy in soul, mind and body, because they will be in harmony with each other.

To reach the ideal man it will require the highest aspiration and the greatest effort. The highest state is beautifully expressed in the following:

"With this awakening and realization one is brought at once *en rapport* with the universe. He feels the power and the thrill of life universal. He goes out from his own little garden spot and mingles with the great universe; and the little perplexities, trials and difficulties of life that today so vex and annoy him, fall away of their own accord by reason of their insignificance. The intuitions become keener and ever more keen and unerring in their guidance. There comes more and more the power of reading men, so that no harm can come from this source. There comes more and more the power of seeing the future, so that more and more true becomes the old adage that coming events cast their shadows before. Health in time takes the place of disease; for all disease and its consequent suffering is merely the result of the violation of law, whether consciously or unconsciously. There comes a spiritual power which, as it is sent out, is adequate for the healing of others the same as in the days of old. The body becomes less gross and heavy, finer in its texture and form, so that it serves far better and responds more readily to the higher impulses of the soul. Matter itself in time responds to the action of these higher forces; and many things that we are accustomed by reason of our limited vision to call miraculous or supernatural become the normal, the natural, the every-day."

The new man will have peace, efficiency and poise.

He is going to think good, act good, and be good, because he will know by doing so he will crowd out the bad. The better material he gives his brain to work on, the better his thoughts will be. Never fear evil; your bad thoughts will do you your greatest harm, so don't let them in. Keep the door always closed to them. There is nothing that can come out of the brain but what went in there. You cannot have good thoughts if you are thinking of bad thoughts.

It is your duty to get yourself right, and always do the best you can, and that is what is expected of you. Don't fret yourself about what others are doing. Don't worry about evils you have no control over.

The Supreme Being is in charge of everything. He will see that all things work out aright. Conditions can be improved only by the individual unit improving. You are one of the tiny spokes in a huge wheel that is gradually turning to perfection.

The coming man will do his part in helping humanity by doing right and acting right.

He will look ever at the present and the future, as he knows the past is dead. You should join the new man brigade and start in to what you know you should do, and keep it up from now on.

He will not go around with a grouch on. He will look pleasant, smile, and speak kindly and cheerily to every one he meets.

He will never do a mean act, and he will never speak a word that is not true about any one, and even if it is the truth and it will hurt some one, he will not say it, but remain quiet.

If you will only live up to the teaching of this book you will be far happier and feel much better when you lay your head down on your pillow at night. Your sleep will be refreshing.

Once more I repeat, do not entertain worry. Direct your attention away from worry thoughts, and keep it on good thoughts. You have a big enough responsibility in your own self, and too much if you try to be responsible for others. You cannot help their troubles by worrying about them. By helping your-

Man and Woman Intended to Be 223

self first, you can show others how to help themselves.

It matters not how hard your life has been, there will be an equal joy for your hardships. There is a joy for every tear you have shed, if you only try to find it. Life will seem better and sweeter when you wear a pleasant face than it will if it is sour.

Never look down upon any one, because you are not perfect. There are people if you were measured up beside, you would not show to the best advantage.

You are better able to understand how to treat others. If you have been without funds and charity has helped you, you will give to charity, because you know what a blessing it is. Every loss, no matter what it is, is an experience which you can turn to your advantage. When you have recovered from your loss you know the only real, actual loss in life is death or loss of health.

The man of the future will not let any one get the best of him, because he will reserve this for himself. Nothing will keep him down, because he will not be downhearted. He will realize that the hard sledging will be the making of him, that he can do more than he thought he could.

He will be greatly helped by understanding the value of charity, justice, smiles; and they will do a great deal to rout his troubles.

He will be larger, broader, fairer, and more tolerant in every way. By his determination to do better he becomes more prosperous and happier while others scold and become duller in every way, and worry themselves into their graves before their time. He will know there is no limit to him, and he should thank the good samaritan that put him on the right road, and be grateful that he had the brains and ability to see the right road.

MY FINAL WORD.

You only need to devote five minutes a day to meditation to make yourself much better. By devoting five minutes at night to absolute solitude you will have wonderful opportunities for mental culture, brain building, and fortifying yourself for the coming day. I want you to get in the habit every day at some convenient time and think of what you have done in the past twenty-four hours that you were awake. Think carefully of every act you have done. The acts you have gone through; the thoughts you have had; the conclusion you arrived at; why you came to that conclusion. Was your solution the right one? By this process you will be calling the mistakes forcibly to your mind and you will be less liable to do them again. The good things will become a regular habit, and those that are not good for you to do, will be just a little harder the next time. In a short while you will avoid them altogether.

Study the lives of any one of our great men of any age, and you will find that he has spent a great deal of his time in solitude, sizing up things and making a mental inventory of what he has been doing, and what he wants to do. Thinking of things that have happened, makes the brain grow and makes it stronger, and you are able to detect and eliminate the things which harm the brain and make it weaker.

Watch for your positive and negative acts. The wise and foolish things you did, see if it would not have been better not to have done them. Are you progressing or going backward in developing good qualities? If the latter is the case get a hustle on yourself and develop the former.

Practice this sizing up for two weeks and notice the improvement. You will be so repaid that you will keep it up.

Analyze your problems and decide for yourself, and act alone. Don't ask advice of others. It weakens your own powers and makes you less certain. Most people will advise you in a way that they know will please you, unless they figure in the results, in which case they will advise you so that they will be benefited. Other people's advice usually is of no value to you.

The strong man lets it alone. Studying this book will give you a great deal of help on how to stand on your own feet. You are the one to solve your problems. Keep this idea before you. Everyone is busy with their own affairs. Everyone should be their own advisor, director and counsellor. Let your will-power dominate and you will be a power and success.

Magnetism is a big subject as you no doubt realize, and a very important one. It is made up of a lot of small things and it is not a mysterious power as some would lead you to believe. It is your duty to yourself first of all, and duty to your loved ones, and to the world to develop this force. There is no greater aim than to want to be invincible and fearless, unconquered and powerful before the world.

The real man will not be affected by the winds and clouds and storms that pass by, but he will be a living, breathing model of God's most wonderful masterpiece, "Man!"

Friends before saying good-bye I wish to impress on your mind that by just reading this work through you have derived some good I hope, but not a great deal. But if you will read it over slowly and get a note book and write down what appeals to your particular needs, and then practice the instruction until it sinks into your subconscious self you will find you will be greatly benefited.

I would advise you only to read one chapter of this book a day, and then think about how much you have read. Write down on paper what you can remember. Then re-read the chapter and see how much you overlooked. This will sustain your interest and get you in the habit of right thinking. If you have been very busy all day and have not had time to do any reading take ten minutes before you go to bed and read a chapter. Think of nothing else but what you are reading. As you lay your head upon your pillow think of the helpful thoughts you have read. Go to sleep thinking of these thoughts.

The next morning after making your toilet, drink two good sized glasses of water and then re-read the chapter you read the night before. After eating your breakfast take up the duties that require your attention for the day, and be determined that you will have the strength and will-power to overcome the things that you have formerly worried and fretted about. If you will follow what you have read in the preceding pages you are going to gain personal magnetism from this time on.

I know the road to lead you over; all I ask for is your sincerity, patience and faith until you know the road. I know I can lead you towards the fountain of strength and to the ladder of progress.

It matters not how many times you have failed in following other systems, I know you can succeed by following mine.

Throughout this work you have been given

cold facts boiled down to the essence. I advocate no isms, freak beliefs or miracles. You can get what you are seeking if you will only put forward the effort. It does you no good to know something if you do not practice it.

Study the lessons one by one, and practice the rules contained therein, patiently, faithfully and persistently and you will acquire a full and better life, and you will become more magnetic.

Great characters are made possible by seeing man as he is and as he ought to be.

FINIS.

BOOK JUNGLE

Bringing Classics to Life

www.bookjungle.com email: sales@bookjungle.com fax: 630-214-0564 mail: Book Jungle PO Box 2226 Champaign, IL 61825

The Two Babylons
Alexander Hislop   QTY

You may be surprised to learn that many traditions of Roman Catholicism in fact don't come from Christ's teachings but from an ancient Babylonian "Mystery" religion that was centered on Nimrod, his wife Semiramis, and a child Tammuz. This book shows how this ancient religion transformed itself as it incorporated Christ into its teachings....

Religion/History   Pages:358
ISBN: *1-59462-010-5*   MSRP *$22.95*

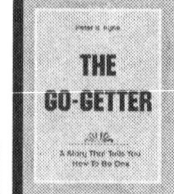

The Go-Getter
Kyne B. Peter   QTY

The Go Getter is the story of William Peck. He was a war veteran and amputee who will not be refused what he wants. Peck not only fights to find employment but continually proves himself more than competent at the many difficult test that are throw his way in the course of his early days with the Ricks Lumber Company...

Business/Self Help/Inspirational   Pages:68
ISBN: *1-59462-186-1*   MSRP *$8.95*

The Power Of Concentration
Theron Q. Dumont

It is of the utmost value to learn how to concentrate. To make the greatest success of anything you must be able to concentrate your entire thought upon the idea you are working on. The person that is able to concentrate utilizes all constructive thoughts and shuts out all destructive ones...

Self Help/Inspirational   Pages:196
ISBN: *1-59462-141-1*   MSRP *$14.95*

Self Mastery
Emile Coue

Emile Coue came up with novel way to improve the lives of people. He was a pharmacist by trade and often saw ailing people. This lead him to develop autosuggestion, a form of self-hypnosis. At the time his theories weren't popular but over the years evidence is mounting that he was indeed right all along...

New Age/Self Help   Pages:98
ISBN: *1-59462-189-6*   MSRP *$7.95*

Rightly Dividing The Word
Clarence Larkin

The "Fundamental Doctrines" of the Christian Faith are clearly outlined in numerous books on Theology, but they are not available to the average reader and were mainly written for students. The Author has made it the work of his ministry to preach the "Fundamental Doctrines." To this end he has aimed to express them in the simplest and clearest manner..

Religion   Pages:352
ISBN: *1-59462-334-1*   MSRP *$23.45*

The Awful Disclosures Of
Maria Monk

"I cannot banish the scenes and characters of this book from my memory. To me it can never appear like an amusing fable, or lose its interest and importance. The story is one which is continually before me, and must return fresh to my mind with painful emotions as long as I live..."

Religion   Pages:232
ISBN: *1-59462-160-8*   MSRP *$17.75*

The Law of Psychic Phenomena
Thomson Jay Hudson

"I do not expect this book to stand upon its literary merits; for if it is unsound in principle, felicity of diction cannot save it, and if sound, homeliness of expression cannot destroy it. My primary object in offering it to the public is to assist in bringing Psychology within the domain of the exact sciences. That this has never been accomplished..."

New Age   Pages:420
ISBN: *1-59462-124-1*   MSRP *$29.95*

As a Man Thinketh
James Allen

"This little volume (the result of meditation and experience) is not intended as an exhaustive treatise on the much-written-upon subject of the power of thought. It is suggestive rather than explanatory, its object being to stimulate men and women to the discovery and perception of the truth that by virtue of the thoughts which they choose and encourage..."

Inspirational/Self Help   Pages:80
ISBN: *1-59462-231-0*   MSRP *$9.45*

Beautiful Joe
Marshall Saunders

When Marshall visited the Moore family in 1892, she discovered Joe, a dog they had nursed back to health from his previous abusive home to live a happy life. So moved was she, that she wrote this classic masterpiece which won accolades and was recognized as a heartwarming symbol for humane animal treatment...

Fiction   Pages:256
ISBN: *1-59462-261-2*   MSRP *$18.45*

The Enchanted April
Elizabeth Von Arnim

It began in a woman's club in London on a February afternoon, an uncomfortable club, and a miserable afternoon when Mrs. Wilkins, who had come down from Hampstead to shop and had lunched at her club, took up The Times from the table in the smoking-room...

Fiction   Pages:368
ISBN: *1-59462-150-0*   MSRP *$23.45*

The Codes Of Hammurabi And Moses - W. W. Davies

The discovery of the Hammurabi Code is one of the greatest achievements of archaeology, and is of paramount interest, not only to the student of the Bible, but also to all those interested in ancient history...

Religion   Pages:132
ISBN: *1-59462-338-4*   MSRP *$12.95*

Holland - The History Of Netherlands
Thomas Colley Grattan

Thomas Grattan was a prestigious writer from Dublin who served as British Consul to the US. Among his works is an authoritative look at the history of Holland. A colorful and interesting look at history....

History/Politics   Pages:408
ISBN: *1-59462-137-3*   MSRP *$26.95*

The Thirty-Six Dramatic Situations
Georges Polti

An incredibly useful guide for aspiring authors and playwrights. This volume categorizes every dramatic situation which could occur in a story and describes them in a list of 36 situations. A great aid to help inspire or formalize the creative writing process...

Self Help/Reference   Pages:204
ISBN: *1-59462-134-9*   MSRP *$15.95*

A Concise Dictionary of Middle English
A. L. Mayhew
Walter W. Skeat

The present work is intended to meet, in some measure, the requirements of those who wish to make some study of Middle-English, and who find a difficulty in obtaining such assistance as will enable them to find out the meanings and etymologies of the words most essential to their purpose...

Reference/History   Pages:332
ISBN: *1-59462-119-5*   MSRP *$29.95*

www.bookjungle.com email: sales@bookjungle.com fax: 630-214-0564 mail: Book Jungle PO Box 2226 Champaign, IL 61825

BOOK JUNGLE

Bringing Classics to Life

www.bookjungle.com email: sales@bookjungle.com fax: 630-214-0564 mail: Book Jungle PO Box 2226 Champaign, IL 61825

The Witch-Cult in Western Europe
Margaret Murray   QTY

The mass of existing material on this subject is so great that I have not attempted to make a survey of the whole of European "Witchcraft" but have confined myself to an intensive study of the cult in Great Britain. In order, however, to obtain a clearer understanding of the ritual and beliefs I have had recourse to French and Flemish sources...

Occult   Pages: 308
ISBN: *1-59462-126-8*   MSRP *$22.45*

Philosophy Of Natural Therapeutics
Henry Lindlahr   QTY

We invite the earnest cooperation in this great work of all those who have awakened to the necessity for more rational living and for radical reform in healing methods...

Health/Philosophy/Self Help   Pages: 552
ISBN: *1-59462-132-2*   MSRP *$34.95*

The Science Of Psychic Healing
Yogi Ramacharaka

This book is not a book of theories it deals with facts. Its author regards the best of theories as but working hypotheses to be used only until better ones present themselves. The "fact" is the principal thing the essential thing to uncover which the tool, theory, is used...

New Age/Health   Pages: 180
ISBN: *1-59462-140-3*   MSRP *$13.95*

A Message to Garcia
Elbert Hubbard

This literary trifle, A Message to Garcia, was written one evening after supper, in a single hour. It was on the Twenty-second of February, Eighteen Hundred Ninety-nine, Washington's Birthday, and we were just going to press with the March Philistine...

New Age/Fiction   Pages: 92
ISBN: *1-59462-144-6*   MSRP *$9.95*

Bible Myths
Thomas Doane

In pursuing the study of the Bible Myths, facts pertaining thereto, in a condensed form, seemed to be greatly needed, and nowhere to be found. Widely scattered through hundreds of ancient and modern volumes, most of the contents of this book may indeed be found; but any previous attempt to trace exclusively the myths and legends...

Religion/History   Pages: 644
ISBN: *1-59462-163-2*   MSRP *$38.95*

The Book of Jasher
Alcuinus Flaccus Albinus

The Book of Jasher is an historical religious volume that many consider as a missing holy book from the Old Testament. Particularly studied by the Church of Later Day Saints and historians, it covers the history of the world from creation until the period of Judges in Israel. It's authenticity is bolstered due to a reference to the Book of Jasher in the Bible in Joshua 10:13

Religion/History   Pages: 276
ISBN: *1-59462-197-7*   MSRP *$18.95*

Tertium Organum
P. D. Ouspensky

A truly mind expanding writing that combines science with mysticism with unprecedented elegance. He presents the world we live in as a multi dimensional world and time as a motion through this world. But this isn't a cold and purely analytical explanation but a masterful presentation filled with similes and analogies...

New Age   Pages: 356
ISBN: *1-59462-205-1*   MSRP *$23.95*

The Titan
Theodore Dreiser

"When Frank Algernon Cowperwood emerged from the Eastern District Penitentiary, in Philadelphia he realized that the old life he had lived in that city since boyhood was ended. His youth was gone, and with it had been lost the great business prospects of his earlier manhood. He must begin again..."

Fiction   Pages: 564
ISBN: *1-59462-220-5*   MSRP *$33.95*

Advance Course in Yogi Philosophy
Yogi Ramacharaka

"The twelve lessons forming this volume were originally issued in the shape of monthly lessons, known as "The Advanced Course in Yogi Philosophy and Oriental Occultism" during a period of twelve months beginning with October, 1904, and ending September, 1905."

Philosophy/Inspirational/Self Help   Pages: 340
ISBN: *1-59462-229-9*   MSRP *$22.95*

Biblical Essays
J. B. Lightfoot

About one-third of the present volume has already seen the light. The opening essay "On the Internal Evidence for the Authenticity and Genuineness of St John's Gospel" was published in the "Expositor" in the early months of 1890, and has been reprinted since...

Religion/History   Pages: 480
ISBN: *1-59462-238-8*   MSRP *$30.95*

Ambassador Morgenthau's Story
Henry Morgenthau

"By this time the American people have probably become convinced that the Germans deliberately planned the conquest of the world. Yet they hesitate to convict on circumstantial evidence and for this reason all eye witnesses to this, the greatest crime in modern history, should volunteer their testimony..."

History   Pages: 472
ISBN: *1-59462-244-2*   MSRP *$29.95*

The Settlement Cook Book
Simon Kander

A legacy from the civil war, this book is a classic "American charity cookbook," which was used for fundraisers starting in Milwaukee. While it has transformed over the years, this printing provides great recipes from American history. Over two million copies have been sold. This volume contains a rich collection of recipes from noted chefs and hostesses of the turn of the century...

How-to   Pages: 472
ISBN: *1-59462-256-6*   MSRP *$29.95*

The Aquarian Gospel of Jesus the Christ
Levi Dowling

A retelling of Jesus' story which tells us what happened during the twenty year gap left by the Bible's New Testament. It tells of his travels to the far-east where he studied with the masters and fought against the rigid caste system. This book has enjoyed a resurgence in modern America and provides spiritual insight with charm. Its influences can be seen throughout the Age of Aquarius.

Religion   Pages: 264
ISBN: *1-59462-321-X*   MSRP *$18.95*

My Life and Work
Henry Ford

Henry Ford revolutionized the world with his implementation of mass production for the Model T automobile. Gain valuable business insight into his life and work with his own auto-biography... "We have only started on our development of our country we have not as yet, with all our talk of wonderful progress, done more than scratch the surface. The progress has been wonderful enough but..."

Biographies/History/Business   Pages: 300
ISBN: *1-59462-198-5*   MSRP *$21.95*

www.bookjungle.com email: sales@bookjungle.com fax: 630-214-0564 mail: Book Jungle PO Box 2226 Champaign, IL 61825

Bringing Classics to Life

BOOK JUNGLE

www.bookjungle.com *email:* sales@bookjungle.com *fax:* 630-214-0564 *mail:* Book Jungle PO Box 2226 Champaign, IL 61825

QTY

☐	**The Rosicrucian Cosmo-Conception Mystic Christianity** *by* **Max Heindel** **ISBN:** *1-59462-188-8* **$38.95** *The Rosicrucian Cosmo-conception is not dogmatic, neither does it appeal to any other authority than the reason of the student. It is, not controversial, but is sent forth in the hope that it may help to clear...* New Age Religion Pages 646
☐	**Abandonment To Divine Providence** *by* **Jean-Pierre de Caussade** **ISBN:** *1-59462-228-0* **$25.95** *"The Rev. Jean Pierre de Caussade was one of the most remarkable spiritual writers of the Society of Jesus in France in the 18th Century. His death took place at Toulouse in 1751. His works have gone through many editions and have been republished...* Inspirational/Religion Pages 400
☐	**Mental Chemistry** *by* **Charles Haanel** **ISBN:** *1-59462-192-6* **$23.95** *Mental Chemistry allows the change of material conditions by combining and appropriately utilizing the power of the mind. Much like applied chemistry creates something new and unique out of careful combinations of chemicals the mastery of mental chemistry...* New Age Pages 354
☐	**The Letters of Robert Browning and Elizabeth Barret Barrett 1845-1846 vol II** **ISBN:** *1-59462-193-4* **$35.95** *by* **Robert Browning** *and* **Elizabeth Barrett** Biographies Pages 596
☐	**Gleanings In Genesis (volume I)** *by* **Arthur W. Pink** **ISBN:** *1-59462-130-6* **$27.45** *Appropriately has Genesis been termed "the seed plot of the Bible" for in it we have, in germ form, almost all of the great doctrines which are afterwards fully developed in the books of Scripture which follow...* Religion/Inspirational Pages 420
☐	**The Master Key** *by* **L. W. de Laurence** **ISBN:** *1-59462-001-6* **$30.95** *In no branch of human knowledge has there been a more lively increase of the spirit of research during the past few years than in the study of Psychology, Concentration and Mental Discipline. The requests for authentic lessons in Thought Control, Mental Discipline and...* New Age/Business Pages 422
☐	**The Lesser Key Of Solomon Goetia** *by* **L. W. de Laurence** **ISBN:** *1-59462-092-X* **$9.95** *This translation of the first book of the "Lemegton" which is now for the first time made accessible to students of Talismanic Magic was done, after careful collation and edition, from numerous Ancient Manuscripts in Hebrew, Latin, and French...* New Age Occult Pages 92
☐	**Rubaiyat Of Omar Khayyam** *by* **Edward Fitzgerald** **ISBN:** *1-59462-332-5* **$13.95** *Edward Fitzgerald, whom the world has already learned, in spite of his own efforts to remain within the shadow of anonymity, to look upon as one of the rarest poets of the century, was born at Bredfield, in Suffolk, on the 31st of March, 1809. He was the third son of John Purcell...* Music Pages 172
☐	**Ancient Law** *by* **Henry Maine** **ISBN:** *1-59462-128-4* **$29.95** *The chief object of the following pages is to indicate some of the earliest ideas of mankind, as they are reflected in Ancient Law, and to point out the relation of those ideas to modern thought.* Religion/History Pages 452
☐	**Far-Away Stories** *by* **William J. Locke** **ISBN:** *1-59462-129-2* **$19.45** *"Good wine needs no bush, but a collection of mixed vintages does. And this book is just such a collection. Some of the stories I do not want to remain buried for ever in the museum files of dead magazine-numbers an author's not unpardonable vanity..."* Fiction Pages 272
☐	**Life of David Crockett** *by* **David Crockett** **ISBN:** *1-59462-250-7* **$27.45** *"Colonel David Crockett was one of the most remarkable men of the times in which he lived. Born in humble life, but gifted with a strong will, an indomitable courage, and unremitting perseverance...* Biographies/New Age Pages 424
☐	**Lip-Reading** *by* **Edward Nitchie** **ISBN:** *1-59462-206-X* **$25.95** *Edward B. Nitchie, founder of the New York School for the Hard of Hearing, now the Nitchie School of Lip-Reading, Inc, wrote "LIP-READING Principles and Practice". The development and perfecting of this meritorious work on lip-reading was an undertaking...* How-to Pages 400
☐	**A Handbook of Suggestive Therapeutics, Applied Hypnotism, Psychic Science** **ISBN:** *1-59462-214-0* **$24.95** *by* **Henry Munro** Health/New Age/Health/Self-help Pages 376
☐	**A Doll's House: and Two Other Plays** *by* **Henrik Ibsen** **ISBN:** *1-59462-112-8* **$19.95** *Henrik Ibsen created this classic when in revolutionary 1848 Rome. Introducing some striking concepts in playwriting for the realist genre, this play has been studied the world over.* Fiction/Classics/Plays 308
☐	**The Light of Asia** *by* **sir Edwin Arnold** **ISBN:** *1-59462-204-3* **$13.95** *In this poetic masterpiece, Edwin Arnold describes the life and teachings of Buddha. The man who was to become known as Buddha to the world was born as Prince Gautama of India but he rejected the worldly riches and abandoned the reigns of power when...* Religion History Biographies Pages 170
☐	**The Complete Works of Guy de Maupassant** *by* **Guy de Maupassant** **ISBN:** *1-59462-157-8* **$16.95** *"For days and days, nights and nights, I had dreamed of that first kiss which was to consecrate our engagement, and I knew not on what spot I should put my lips..."* Fiction/Classics Pages 240
☐	**The Art of Cross-Examination** *by* **Francis L. Wellman** **ISBN:** *1-59462-309-0* **$26.95** *Written by a renowned trial lawyer, Wellman imparts his experience and uses case studies to explain how to use psychology to extract desired information through questioning.* How-to/Science Reference Pages 408
☐	**Answered or Unanswered?** *by* **Louisa Vaughan** **ISBN:** *1-59462-248-5* **$10.95** *Miracles of Faith in China* Religion Pages 112
☐	**The Edinburgh Lectures on Mental Science (1909)** *by* **Thomas** **ISBN:** *1-59462-008-3* **$11.95** *This book contains the substance of a course of lectures recently given by the writer in the Queen Street Hall, Edinburgh. Its purpose is to indicate the Natural Principles governing the relation between Mental Action and Material Conditions...* New Age Psychology Pages 148
☐	**Ayesha** *by* **H. Rider Haggard** **ISBN:** *1-59462-301-5* **$24.95** *Verily and indeed it is the unexpected that happens! Probably if there was one person upon the earth from whom the Editor of this, and of a certain previous history, did not expect to hear again...* Classics Pages 380
☐	**Ayala's Angel** *by* **Anthony Trollope** **ISBN:** *1-59462-352-X* **$29.95** *The two girls were both pretty, but Lucy who was twenty-one who supposed to be simple and comparatively unattractive, whereas Ayala was credited, as her Bombwhat romantic name might show, with poetic charm and a taste for romance. Ayala when her father died was nineteen...* Fiction Pages 484
☐	**The American Commonwealth** *by* **James Bryce** **ISBN:** *1-59462-286-8* **$34.45** *An interpretation of American democratic political theory. It examines political mechanics and society from the perspective of Scotsman James Bryce* Politics Pages 572
☐	**Stories of the Pilgrims** *by* **Margaret P. Pumphrey** **ISBN:** *1-59462-116-0* **$17.95** *This book explores pilgrims religious oppression in England as well as their escape to Holland and eventual crossing to America on the Mayflower, and their early days in New England...* History Pages 268

www.bookjungle.com *email:* sales@bookjungle.com *fax:* 630-214-0564 *mail:* Book Jungle PO Box 2226 Champaign, IL 61825

BOOK JUNGLE

Bringing Classics to Life

www.bookjungle.com *email:* sales@bookjungle.com *fax:* 630-214-0564 *mail:* Book Jungle PO Box 2226 Champaign, IL 61825

			QTY
The Fasting Cure *by Sinclair Upton*	ISBN: *1-59462-222-1*	$13.95	☐
In the Cosmopolitan Magazine for May, 1910, and in the Contemporary Review (London) for April, 1910, I published an article dealing with my experiences in fasting. I have written a great many magazine articles, but never one which attracted so much attention...		*New Age/Self Help/Health Pages 164*	
Hebrew Astrology *by Sepharial*	ISBN: *1-59462-308-2*	$13.45	☐
In these days of advanced thinking it is a matter of common observation that we have left many of the old landmarks behind and that we are now pressing forward to greater heights and to a wider horizon than that which represented the mind-content of our progenitors...		*Astrology Pages 144*	
Thought Vibration or The Law of Attraction in the Thought World	ISBN: *1-59462-127-6*	$12.95	☐
by William Walker Atkinson		*Psychology/Religion Pages 144*	
Optimism *by Helen Keller*	ISBN: *1-59462-108-X*	$15.95	☐
Helen Keller was blind, deaf, and mute since 19 months old, yet famously learned how to overcome these handicaps, communicate with the world, and spread her lectures promoting optimism. An inspiring read for everyone...		*Biographies/Inspirational Pages 84*	
Sara Crewe *by Frances Burnett*	ISBN: *1-59462-360-0*	$9.45	☐
In the first place, Miss Minchin lived in London. Her home was a large, dull, tall one, in a large, dull square, where all the houses were alike, and all the sparrows were alike, and where all the door-knockers made the same heavy sound...		*Childrens/Classic Pages 88*	
The Autobiography of Benjamin Franklin *by Benjamin Franklin*	ISBN: *1-59462-135-7*	$24.95	☐
The Autobiography of Benjamin Franklin has probably been more extensively read than any other American historical work, and no other book of its kind has had such ups and downs of fortune. Franklin lived for many years in England, where he was agent...		*Biographies/History Pages 332*	

Name	
Email	
Telephone	
Address	
City, State ZIP	

☐ Credit Card   ☐ Check / Money Order

Credit Card Number	
Expiration Date	
Signature	

Please Mail to: Book Jungle
               PO Box 2226
               Champaign, IL 61825
or Fax to: 630-214-0564

ORDERING INFORMATION

web: *www.bookjungle.com*
email: *sales@bookjungle.com*
fax: *630-214-0564*
mail: *Book Jungle PO Box 2226 Champaign, IL 61825*
or PayPal *to sales@bookjungle.com*

Please contact us for bulk discounts

DIRECT-ORDER TERMS

20% Discount if You Order Two or More Books
Free Domestic Shipping!
Accepted: Master Card, Visa, Discover, American Express

www.ingramcontent.com/pod-product-compliance
Lightning Source LLC
Chambersburg PA
CBHW081220170426
43198CB00017B/2672